Worldmaking and Cuneiform Antiquity

Objects of knowledge exist within material, immaterial, and conceptual worlds. Once the world is conceived from the perspective of others, the physical ontology of modern science no longer functions as a standard by which to understand other orderings of reality, whether from ethnographical or historical sources. Because premodern and non-Western sources attest to a plurality of sciences practiced in accordance with different ways of worldmaking from that of the modern West, their study belongs to the history of science, the philosophy of science, and the sociology of science, as well as the anthropology of science. In *Worldmaking and Cuneiform Antiquity*, Francesca Rochberg extends an anthropology of science to the historical world of cuneiform texts of ancient Babylonia. Exploring how Babylonian science has been understood, she proposes a new direction for scholarship by recognizing the world of ancient science, not as a less developed form of modern science, but as legitimate and real in its own right.

Francesca Rochberg is Catherine and William L. Magistretti Distinguished Professor of Near Eastern Studies Emerita at the University of California, Berkeley. A John D. and Catherine T. MacArthur fellow and John Simon Guggenheim fellow, she was elected to the American Philosophical Society in 2008. Rochberg is the author of *The Heavenly Writing: Divination, Horoscopy, and Astronomy in Mesopotamian Culture* and *Before Nature: Cuneiform Knowledge and the History of Science.*

Worldmaking and Cuneiform Antiquity

An Anthropology of Science

FRANCESCA ROCHBERG
University of California, Berkeley

CAMBRIDGE
UNIVERSITY PRESS

CAMBRIDGE
UNIVERSITY PRESS

Shaftesbury Road, Cambridge CB2 8EA, United Kingdom

One Liberty Plaza, 20th Floor, New York, NY 10006, USA

477 Williamstown Road, Port Melbourne, VIC 3207, Australia

314–321, 3rd Floor, Plot 3, Splendor Forum, Jasola District Centre, New Delhi – 110025, India

103 Penang Road, #05–06/07, Visioncrest Commercial, Singapore 238467

Cambridge University Press is part of Cambridge University Press & Assessment, a department of the University of Cambridge.

We share the University's mission to contribute to society through the pursuit of education, learning and research at the highest international levels of excellence.

www.cambridge.org
Information on this title: www.cambridge.org/9781009522335

DOI: 10.1017/9781009522298

When citing this work, please include a reference to the DOI 10.1017/9781009522298

First published 2024

A catalogue record for this publication is available from the British Library.

Library of Congress Cataloging-in-Publication Data
NAMES: Rochberg, Francesca, 1952– author.
TITLE: Worldmaking and cuneiform antiquity : an anthropology of science / Francesca Rochberg, University of California, Berkeley.
DESCRIPTION: Cambridge, United Kingdom ; New York, NY : Cambridge University Press, 2024. | Includes bibliographical references and index.
IDENTIFIERS: LCCN 2024010710 | ISBN 9781009522335 (hardback) | ISBN 9781009522281 (paperback) | ISBN 9781009522298 (ebook)
SUBJECTS: LCSH: Science, Ancient. | Assyriology. | Astrology, Assyro-Babylonian. | Astronomy, Assyro-Babylonian. | Cuneiform tablets – Iraq – Babylonia. | Cuneiform tablets – Assyria. | Scientific literature – Iraq – Babylonia – History and criticism. | Scientific literature – Assyria – History and criticism.
CLASSIFICATION: LCC Q124.95 .R63 2024 | DDC 509.3–dc23/eng/20240506
LC record available at https://lccn.loc.gov/2024010710

ISBN 978-1-009-52233-5 Hardback

Dedicated to the memory of my mother,
Gene Rosenfeld Rochberg
1919–2016
Love eterne

ina amat Anu u Antu lišlim

Can a world that alters with time and from one community to the next correspond to what is generally referred to as "the real world"? I do not see how its right to that title can be denied.

Thomas S. Kuhn

Contents

Figures

Preface

It would be easy to begin an inventory of the phenomena belonging to our various modern fields of scientific inquiry. In no particular order, we could say chemical elements, supernovae, molecules and quarks, biological organisms, minerals, climate and weather, and many other things. Even when we ourselves do not fully understand what some of these phenomena are, we are nevertheless on familiar ground in that we have no difficulty recognizing them as belonging to our world of scientific inquiry. What about the phenomena belonging to the ancient cuneiform scribes' world of inquiry, or indeed some other premodern Western or non-Western community of inquirers? In the case of cuneiform antiquity and in no particular order, we could say stars, the periodicity of eclipses, the heliacal rising of Venus, symptoms of disease, magical preparations for cures, stones that are good for amulets, demons and divine agents of causality, the color of a planet, the length of the year, the meaning of an eclipse that sets before it clears, the appearance of the intestinal coils of a sacrificed lamb, the meaning of a mole on someone's chin, winds and clouds, the water level of the Euphrates, and many other things. Instantly, it is clear that the phenomena of interest to the ancient scribes of Babylonia and Assyria, the objects of their inquiry, description, and classification, have a fundamentally different aspect, and we find ourselves on unfamiliar ground. Even where there are apparent overlaps, such as in the study of the stars, we are struck by deep-seated differences in why the stars were of interest to the ancients and what methods they devised to understand and describe them.

Not only were the reasons for and methods of inquiry into phenomena in cuneiform antiquity fundamentally different from our own, or from those of other communities of inquirers in other times and places, but the phenomenal world of cuneiform scribal knowledge was also a world answering to their questions and their reasons for the observing, systematizing, describing, and analyzing of phenomena that we recognize as the cuneiform sciences. In short, the sciences of cuneiform antiquity were oriented toward a world utterly different from our own. How to analyze and understand that difference as a function of history and culture is the crux of this book.

To that end, the approach of the present work is twofold: the first is to examine issues in the historiography of premodern science, principally but not exclusively in cuneiform antiquity, and second is to consider the methodological possibility of infusing the historiography of science with the cultural concerns of an anthropology of science. For clarity, this dual focus is presented in two parts. The integration of historiography of science (Part I) with anthropology (Part II) hinges on the attention to cultural dimensions of scientific knowledge and scientific practice in the distant past.

In order to bring the two together and show their interdependence, one of the central claims of this book is that scientific epistemologies (systems and methodologies of knowledge about the world's phenomena) are inextricably bound up with ontologies (ideas about what phenomena there are in the world). Accordingly, knowledge is inseparable from the nature and classification of the things that are known, as well as the very conceptual framework – reflected explicitly in ideas or implicitly in practices – within which things are known.

I do not propose to exchange the epistemological for the ontological in an inquiry into the history of science, but to see these as inseparable aspects of worldmaking discernible in cuneiform sources in which we find descriptions of the world, its parts, and its phenomena. Although the word "worldmaking" (or world-making) is attested as far back as the eighteenth century and has a somewhat wider usage now (e.g., Kondo 2018; Getachew 2019), the sense in which I use it is associated with philosopher Nelson Goodman's 1978 *Ways of Worldmaking*. In that work, he developed an argument for the relation of the notion of "world," in the sense of the way(s) the world is, to constructional frameworks of description and conceptualization. Importantly, and already articulated by him years before the 1978 book, Goodman's notion was that there is no one way the world is, but many ways, captured in many constructed world-versions (Goodman 1960). The present book aims to exploit this idea for historical epistemologies, ontologies, and the historiography of early science.

The historicity of knowledge is well accepted in the history of science today. Ontology, in the sense of "what there is," as the inseparable complement of epistemology, in the sense of "what is known," has less often been historicized. When both epistemologies and ontologies are historicized, the view of unique truth and a singular world fixed by the ontological realm we define as nature gives rise to a diversity of real yet possibly conflicting ways the world can be made. I refer here specifically to premodern historical contexts, both Western and non-Western, but the same holds for non-Western anthropological and ethnographic contexts.

Until the latter part of the last century, the premodern and non-Western forms of cuneiform knowledge were classified as science only when they were found to resemble or resonate with Western scientific ideas and practices. What is at stake in this book is not the question of how like or unlike later sciences the ancient traditions were. It is rather to advocate for a more culturally informed approach to the historiography of cuneiform sciences, an approach that takes a page from the field of anthropology, specifically from the attention paid in recent anthropological research to ontologies and the question of worlds.

The understanding of science as a historical phenomenon, not only of the premodern Western but also of the truly *other* sciences, such as the diverse non-Western systems of knowledge around the globe, warrants further consideration and greater clarification. My remit is not as extensive as that. I will focus more sharply on the sciences of Babylonia and Assyria than on other parts of Middle Eastern antiquity (such as Egypt, Persia, or Judea). I am, however, not aiming for complete coverage of all possibly relevant sources even within the cuneiform world, but rather for exemplification of aspects of cuneiform knowledge in order to expand our consideration of this evidence in new directions.

Augmenting our historiographical method with a cultural focus has an intrinsic relationship to anthropology in that discipline's concern with cultures and cultural ways of perceiving, experiencing, and imagining diverse inhabited worlds. My reflections on worldmaking, however, are directed at historical sources for science, cuneiform science in particular, and the relation of the epistemological to the ontological in that context. They are the result of my own engagement not only with cuneiform primary sources but equally with the philosophy of science concerned with the issue of ontological plurality and relativity from Goodman's time until today, and the responses of anthropologists to questions of ontology.

Worldmaking and Cuneiform Antiquity stems from a desire to push modern methods in the history of cuneiform science into a new realm of ideas about the relationship of science to cultures and the interaction of the members of those cultures of science with their worlds. The project to understand the cuneiform sciences began far upstream when (1) in the late nineteenth to mid twentieth century, the work of Joseph Epping, Franz Xaver Kugler, and then Otto Neugebauer and his contemporaries introduced Babylonian mathematical astronomy to the history of the exact sciences (mathematics and mathematical astronomy) but preserved, by virtue of its being the prevailing assumption of their day, the single world of nature as the object of science; it continued when (2) in the 1970s to 1990s, David Pingree introduced the nonexact sciences, primarily astrology and magic (in Babylonian, Indian, and Classical texts) as appropriate sources for understanding early sciences and exchanged the one culture of science for the many cultures of science; this was followed by (3) the early twenty-first-century work of Eleanor Robson, Mathieu Ossendrijver, and others who adopted a sociology of science and a network approach restoring the cuneiform sciences to the social contexts within which they emerged. Downstream of these important contributions to our knowledge of science in cuneiform texts, we come to the present work in which I advance the idea that (4) an anthropology of science together with worldmaking and historical ontologies can profitably add to existing methodologies of history, philosophy, and sociology of science in premodern contexts. Drawing upon recent developments in the so-called ontological turn within anthropology affords the exchange of the concept of the many cultures of science as takes on the single external world (nature) for that of many diverse sciences as systems of knowledge of many diverse phenomenal worlds.

The challenge taken up in this book is to move away from historical analyses based on the assumption of the one world of modern science with its unique truth and fixed discoverable world of nature in order to understand more directly the diversity and multiplicity of worlds described in historical sources. The anthropology of premodern science, in emphasizing the cultural content of scientific texts and practices, seems to me an important addition to our existing historiographical methodologies, not only for the cuneiform sciences, but also for the study of non-Western and premodern science more generally.

Acknowledgments

I owe a great debt of gratitude to my colleagues, students, and friends from Berkeley and outside Berkeley, from both the Assyriological and the history of science communities, who have supported me in many ways. They are Eduardo Escobar, Hassan Rezakhany, Jay Crisostomo, Niek Veldhuis, Laurie Pearce, Rita Lucarelli, Ron Hendel, Jerry Cooper, Carol Bier, Carolyn Merchant, Aaron Tugendhaft, Gil Breger, Daniel Boyarin, Mark Geller, Irving Finkel, Ann Guinan, Abraham Winitzer, Cale Johnson, Alan Bowen, Liba Taub, Mathieu Ossendrijver, John Steele, Lis Brack-Bernsen, Geoffrey Lloyd, Willard McCarty, Aparecida Villaça, Sonja Brentjes, and Florentina Badalanova Geller.

Although parts of this book trace back many years in my thinking, the manuscript began to crystallize during a visiting scholar's stay in 2018 at the Max Planck Institute for the History of Science (MPIWG) in Berlin, where interaction with other scholars pursuing projects to do with *Wissensgeschichte* across the premodern non-Western world was critical in starting this project. To Lorraine Daston, I owe huge debt of gratitude for this opportunity. Of more recent influence on my thinking are social anthropologists Willard McCarty and Aparecida Villaça, who extended the invitation to participate in Science of the Forest, Science of the Past (SFSP) II, spearheaded by Sir Geoffrey Lloyd. I thank them all.

Thanks are also due the Institute for the History of Knowledge of Antiquity at the Free University, Berlin for hosting me in the summer of 2022 and giving me the opportunity to field some of the ideas of this book. For that, I am especially indebted to J. Cale Johnson, director, and Mathieu Ossendrijver, head of ZODIAC – Ancient Astral Science in Transformation. I also thank Eva Cancik-Kirschbaum at the Free

University, Berlin for her intellectual support and encouragement. I am also grateful to Professors Helge Kragh of the Faraday Institute at Aarhus University and Kirsti Andersen of Aarhus University for their generous help with information concerning Olaf Pedersen.

I have used parts of previously published material in this work, including "Beyond Binarism in Babylon," in "History and Human Nature," ed. Willard McCarty and Brad Inwood, special issue, *Interdisciplinary Science Reviews* 35, nos. 3–4 (2010): 253–65; "These Bones Live!" *KNOW: A Journal on the Formation of Knowledge* 1 (2017): 67–83; "The Catalogues of Enūma Anu Enlil," in *Assyrian and Babylonian Scholarly Text Catalogues: Medicine, Magic and Divination*, ed. Ulrike Steinert, 119–34 (Boston: De Gruyter, 2018); "The Anthropology of Science: The Cuneiform World," *JNES* 78 (2019): 253–71; "The Restoration of Ancient Cuneiform Knowledge and Its Implications," in *Wissensoikonomien: Ordnung und Transgression vormoderner Kulturen*, ed. Nora Schmidt, Nikolas Pissis, and Gyburg Uhlmann, 151–60 (Wiesbaden: Harrassowitz, 2021); "Mereological Themes in Cuneiform Worldmaking," in "Science in the Forest, Science in the Past Pt. 2," ed. Willard McCarty, G. E. R. Lloyd, and Aparecida Vilaça, special issue, *Interdisciplinary Sciences Reviews* 46, no. 3 (2021): 386–404. Reprinted in *Science in the Forest, Science in the Past: Further Interdisciplinary Explorations* (London: Routledge, 2022); "*Ṭupšarrūtu* and the Historiography of Science," *Claroscuro* 2 (2021): 1–27; "Imago Mundi: Worldmaking, Science, and the Ontological Turn," *Mêtis* 20 (2022): 17–33.

I thank the Trustees of the British Museum for permission to publish the photographs of the Stone Prism of Esarhaddon and the tablet BM 34580, and Editions Geuthner for permission to use an extract from René Labat, *Manuel d'épigraphie akkadienne* (1948, 1988 6th edition).

Gratitude and appreciation are also due to Kelly Burch, who formatted the manuscript in its earliest stage, made the index, and helped in inestimable ways. Thanks are also due the anonymous readers of the manuscript for their astute reviews and suggestions for revision. I have tried my best to take their recommendations on board. Last but not least, I want to acknowledge my editor Beatrice Rehl, who gave me substantive input and guidance together with ultimate freedom. I thank her for her steadfast support.

Notes on Chronological and Typographical Conventions

CHRONOLOGY

For periodization of historical periods for the region of the ancient Middle East, see https://cdli.ox.ac.uk/wiki/doku.php?id=adopted_periodisation_in_cdli. The following list is limited to the chronological periods cited in this work. They are for the most part periods designated by terms of political rule. In the case of the Late Babylonian period, however, Late Babylonian refers to cuneiform texts of the second half of the first millennium BCE not datable to any particular dynastic period.

Ur III ca. 2100–2000 BCE
Old Babylonian period ca. 2000–1600 BCE
Neo-Assyrian period 911–612 BCE
Neo-Babylonian period 625–539 BCE
Late Babylonian period ca. 500 BCE–1st century CE
Achaemenid period 547–331 BCE
Seleucid period 312–63 BCE
Hellenistic period 323–32 BCE
Arsacid period 247 BCE–224 CE
Sasanian period 224 BCE–651 CE
Late Antiquity late 3rd century–7th century CE
Middle Ages 5th–15th century CE
Renaissance ca. 1450–1650 CE

TYPOGRAPHICAL CONVENTIONS

Names and words taken from cuneiform texts are spelled in accordance with transliteration and transcription conventions as follows: Sumerian is rendered in lowercase Roman separated by periods, for example, ú.ebur "spring grass." Sumerian logograms used in Akkadian are rendered in uppercase Roman separated by periods, for example, UB.DA.LIMMU_2.BA "the four quarters." Superscripts, e.g., ^{LÚ}ummânu, indicate Sumerian determinatives used in the writing of Akkadian words to mark a classification; in this case, LÚ "man" classifies the word ummânu as a profession. Superscripted determinatives can be pre- or postposed.

Akkadian is rendered in italics, e.g., *kippat erbetti* "circle of the four quarters."

[ ] around Akkadian transcription represents restored readings due to breaks in the tablet.

Numbers in the sexagesimal system (base 60) are rendered with a comma separating sexagesimal places and a semicolon separating integers from fractions, for example, $6,30 = (6 \times 60) + 30 = 390$, but $6;30 = 6 + (30 \div 60) = 6.5$.

Glossary of Astronomical Terminology Cited

The following is meant as a guide to the terms cited in this book, not as a complete glossary of astronomical terms. For further terminology, see Bowen and Rochberg 2020, 631–52, "Historical Glossary of Important Terms in Hellenistic Astronomy."

GENERAL

Celestial equator: The equinoctial circle at right angles to the axis of rotation of the celestial sphere detectable by the motion of the fixed stars. It lies in the plane of the earth's own equator. This circle is the basis of the coordinate system of right ascension and declination.

Ecliptic: The oblique circular path defined by the Sun's eastward motion against the stars (zodiacal circle). This circle is the basis of the coordinate system of celestial longitude and latitude.

Synodic period: The period of a celestial body's return to a position relative to the sun. For example, the synodic month brings the return of the moon to its conjunction with the sun and takes between 29.18 days and 29.93 days.

Sidereal period: The period of a celestial body's return to a position relative to the fixed stars. For example, the sidereal year (in which the sun describes 360°, a period of 365.256366 days), the sidereal month (in which the moon returns to its position with respect to the stars, a period of 27.321 days), or the sidereal period of a planet in which an individual planet returns to an initial position relative to the stars.

Right ascension: Right ascension is measured eastward along the celestial equator (equinoctial circle) in degrees from 0° to 360° beginning at the Vernal Equinoctial Point (vernal equinox).
Declination: Declination is measured from the plane of the equinoctial circle along circles at right angles to the Celestial Poles.
Celestial longitude: Celestial longitude measures the position of a celestial body in degrees along the ecliptic (zodiacal circle) eastward from 0° to 360° starting from the Vernal Equinoctial Point (vernal equinox).
See Aaboe 2001, 1–5.

GREEK ASTRONOMY

Deferent: The circle of the celestial sphere upon the circumference of which the center of a planet's epicycle revolves uniformly.
Epicycle: The circle upon which the planet rotates eastward and the center of which rotates uniformly, that is with constant angular rate. The center of uniform motion need not be the center of the deferent.
Eccentric: A circle that does not have the center of the cosmos or equinoctial circle as it center.
See Evans 2020, 112–24.

BABYLONIAN ASTRONOMY

Linear zigzag function: A continuous periodic function used in System B derived from the values (y_n) of an astronomical table where the values of y_n represent event numbers (n = 1, 2, . . .). A value y_n is found from the preceding value (y_{n-1}) by a constant difference line by line (+/–d) between two extrema, a maximum (M) and a minimum (m), that define an amplitude of Δ. The parameters of a linear zigzag function are M, m, Δ, and d. The period (P) will be found as

$$P = \frac{2\Delta}{d}.$$

Step function: A discontinuous periodic function used in System A derived from the values (y_n) of an astronomical table where the values of y_n represent celestial longitudes (λ) that proceed with constant rate (ω) within a zone, or arc (α) delimited by a jumping point, after which point the difference ($y_{n+1} - y_n$) is modified by a ratio specific to the celestial body in question. The rates (ω) of progress in longitude are found by the differences in longitude ($\Delta\lambda$) along the zodiacal circle (ecliptic). The sum of the arcs (α) is 360°, i.e., $\alpha_1 + \alpha_2 + \ldots + \alpha_n = 360°$. The period (P) is found as $P = \frac{\alpha_1}{\omega_1} + \frac{\alpha_2}{\omega_2} + \ldots + \frac{\alpha_n}{\omega_n}$. The period (P) can also be defined as the number of synodic phenomena that fit within the 360° of the ecliptic, thus $P = \frac{6,0}{\Delta\lambda} = \frac{\Pi}{Z}$ where Π is the number of phenomena and Z is the number of sidereal rotations.

See Neugebauer 1955, 28–32 and figs. 6–7; Ossendrijver 2012, 42–53, Bowen and Rochberg 2020, 651–52.

Abbreviations

4R	H. C. Rawlinson, *The Cuneiform Inscriptions of Western Asia*. Vol. IV: *A Selection from the Miscellaneous Inscriptions of Assyria* (London: The British Museum, 1875).
ACT	Otto Neugebauer, *Astronomical Cuneiform Texts* (London: Lund Humphries, 1955).
ADART	Abraham J. Sachs and Hermann Hunger, *Astronomical Diaries and Related Texts from Babylonia*. Vol. 1: *Diaries from 652 BC to 262 BC* (Vienna: Österreichische Akademie der Wissenschaften, 1988); Abraham J. Sachs and Hermann Hunger, *Astronomical Diaries and Related Texts from Babylon*. Vol. 2: *Diaries from 261 bc to 165* BC (Vienna: Österreichische Akademie der Wissenschaften, 1989); Abraham J. Sachs and Hermann Hunger, *Astronomical Diaries and Related Texts from Babylon*. Vol. 3: *Diaries from 164* BC *to 61* BC (Vienna: Österreichische Akademie der Wissenschaften, 1996); Hermann Hunger, including materials by Abraham J. Sachs, *Astronomical Diaries and Related Texts from Babylonia*. Vol. 4: *Undated Diaries and Addenda* (Vienna: Österreichische Akademie der Wissenschaften, 2022); Hermann Hunger and Abraham J. Sachs, *Astronomical Diaries and Related Texts from Babylonia*. Vol. 5: *Lunar and Planetary Texts*, with an Appendix by John M. Steele (Vienna: Österreichische Akademie der Wissenschaften, 2001); Hermann Hunger, including materials by Abraham

J. Sachs, *Astronomical Diaries and Related Texts from Babylon*. Vol. 6: *Goal Year Texts* (Vienna: Österreichische Akademie der Wissenschaften, 2006); Hermann Hunger, including materials by Abraham J. SachsVol. 7: *Almanacs and Normal Star Almanacs* (Vienna: Österreichische Akademie der Wissenschaften, 2014).

AO — Museum siglum of the Louvre, Paris (Antiquités orientales).

BAM — Franz Köcher, *Die babylonisch-assyrische Medizin in Texten und Untersuchungen* (Berlin: De Gruyter, 1963–).

BM — Museum siglum of the British Museum, London (British Museum).

CAD — *The Chicago Assyrian Dictionary*, vols. 1–21 (Chicago: University of Chicago, Oriental Institute, 1956–2010).

CCAG — Franz Cumont, Franz Boll, and Wilhelm Kroll, eds., *Catalogus Codicum Astrologorum Graecorum*. 12 vols. (Brussels: Lamertin, 1898–1953).

CT — *Cuneiform Texts from Babylonian Tablets in the British Museum* (London: British Museum Publications, 1896–).

DCCLT — Digital Corpus of Cuneiform Lexical Texts.

ePSD — electronic Pennsylvania Sumerian Dictionary Project (http://psd.museum.upenn.edu/epsd/index.html).

ETCSL — *The Electronic Text Corpus of Sumerian Literature* (Oxford: Oxford University, Oriental Institute, 1997–2006).

HAMA — Otto Neugebauer, *A History of Ancient Mathematical Astronomy*, 3 vols. (Berlin: Springer Verlag, 1975).

KAR — Erich Ebeling, *Keilschrifttexte aus Assur religiösen Inhalts* I/II (= WVDOG 28, 1919; 34, 1923).

LBAT — Abraham J. Sachs, *Late Babylonian Astronomical and Related Texts Copied by T. G. Pinches and J. N. Strassmaier*. With the cooperation of J. Schaumberger (Providence, RI: Brown University Press, 1955).

RAcc. — François Thureau-Dangin, *Rituals accadiens* (Paris: Editions E. Leroux, 1921).

Rm. — Museum siglum of the British Museum, London (Rassam).

SpTU 4 — Egbert von Weiher, *Spätbabylonische Texte aus Uruk*, vol. IV (Mainz am Rhein: Gebr. Mann Verlag, 1993).

TCL	*Textes cunéiformes du Louvre* (Paris: Musées du Louvre, 1910–).
VAB	*Vorderasiatische Bibliothek* (Leipzig: Hinrichs, 1907–).
VAT	Museum siglum of the Vorderasiatisches Museum, Berlin (Vorderasiatische Abteilung. Tontafeln).
YOS	Yale Oriental Series, Babylonian Texts (New Haven, CT: Yale University Press, 1915–).

Introduction

> One culture is not a less perfect version of another; winter is not a rudimentary spring; summer is not an undeveloped autumn.
>
> – Isaiah Berlin (1982, 108)

OLD AND NEW PATHS

The material of central interest to this book belongs to the ancient historical world of the Middle East, specifically to what I have previously and continue to classify as science in Assyro-Babylonian cuneiform texts. Cuneiform texts were my point of departure, but they led me down an unexpected path for the historiography of premodern science in the direction of the anthropology of science. As a consequence, in contrast to previous emphases on historical epistemologies reflected in historical bodies of knowledge, my interest shifts here to historical ontologies reflected in the worlds of the scribal compilers and practitioners of those bodies of knowledge. The shift in emphasis toward ontology, I will argue, can be theorized in terms of worldmaking, best approached by means of an anthropology of premodern science.

There is currently an array of analytical approaches to the nature of science: the mostly epistemological emphasis of the philosophy of science, the macroanalytic social structural or institutional emphasis of the sociology of science, the historical emphasis of the history of science, and the cultural emphasis of the anthropology of science. The history and philosophy of science are regularly combined, as is the sociology of science with either or both. Of the four, the anthropology of science does not have the foothold on the discourse as do the history, philosophy, or sociology of science. Whereas the anthropology of science still occupies a marginal place on the map of the various discourses about science, its cultural focus has much to offer for the study of historical and non-Western, and thus cuneiform, sciences.

An anthropology of science oriented not to modern Western scientific practices but rather to non-Western and premodern scientific – or knowledge – cultures, whose ideas stem from ontologies and epistemologies that require translation on many levels in order to be comprehensible, or to be at all comparable with more familiar ways of knowing, faces three immediate problems. The first is how to define or classify a distinctly "other" historical system of thought without using the old criteria establishing a demarcation between science and nonscience and whose primary reference was the modern Western world. The second is how to understand another system and the "other world" implied by its alien conceptual frameworks and schemes, in other words how to overcome so-called incommensurability and make those other systems, and the worlds they represent, translatable to our own terms, preferably with minimal distortion. The third is the problem of conceptual relativism posed by different and distinct constructional systems, classificatory frameworks, or conceptual schemes.

Each of these problems, namely demarcation, incommensurability, and relativism, generated a considerable amount of interest in the philosophy of science community through the twentieth century: demarcation with respect to science versus nonscience (Popper [1935] 2002); incommensurability with respect to different scientific "paradigms" or "disciplinary matrices" for theory (as the Ptolemaic versus the Copernican planetary theories exemplified for Kuhn 1957, [1962] 1996); and conceptual relativism with respect to the perceived damage done to absolute truth and reality if science is socially or even cognitively constructed (in favor of relativism, with certain constraints, were Goodman [1951] 1977, 1960, 1975 and Quine 1968, 1969, 1992; against was Davidson 1973, 2005). Although these are subjects well suited to a history of the history of science, they are not mentioned here as historical curiosities, no longer relevant. I am aware that moribund assessments have appeared about all three: in 1983, Larry Laudan proclaimed "the demise" of the demarcation problem (1983a, 1983b); in 2016, Lorraine Daston (2016, 119), as a commentary on contemporary tastes among members of the field of the history of science, said relativism (as well as incommensurability and irrationality) was a worn-out theme, passé for historians of (modern) science; finally, in 2018, Howard Sankey visited the grave of Kuhnian incommensurability, where he saw grass growing on it.[1]

Despite these well-published funerals for older ways of framing what are essentially problems of difference, the issue of difference continues to burn across many disciplines. By its very nature, anthropology operates in the realm of plurality and difference. Clifford Geertz summed up the problem faced by anthropologists as "how to penetrate a form of life not merely different from but incompatible with their own" (Geertz 2010, 17). In the Geertzian mode, Martin Holbraad (2012, xv) more recently defined "anthropology's flagship problem" as that of making sense of "others." I submit that with regard to cuneiform sciences, we are equally faced with fundamental differences in need of explication.

To make the complex landscape of these intellectual discourses (history and anthropology of science) clear so as to bring their respective emphases to bear on the cuneiform sciences necessitates the division of this book in two parts. My aim is, first, to cast a look back in Part I in order to explore the historiographical underpinnings of modern scholarship about cuneiform sciences together with its implicit philosophical commitments. My interest is partly the astronomical sciences, using that term to encompass descriptive, schematic, and mathematical astronomy as well as celestial divination and horoscopy. The astronomical sciences, however, were part of a larger ancient intellectual culture whose nature vis-à-vis science warrants consideration. Second, I set out on a less well-explored path to view the dimensions of science found in that broader culture. When we concern ourselves not only with historical knowledge and methods of knowing but also with the objects of knowledge and how those objects existed within worlds, material, immaterial, and conceptual, we move in Part II into the terrain of the anthropology of science. I view this exploration as belonging to the anthropology of science because its central question is the relationship a culture strikes with its own world. In ancient Babylonia, we are afforded access to this relationship almost entirely through its textual traditions.

This book, therefore, considers both old and new paths to the historiography of the Babylonian systems of knowledge and of the ways its

[1] Sankey 2018, 87: "Returning to the topic from the perspective of the contemporary scene in the philosophy of science is like visiting a battlefield from a forgotten war. The positions of the warring sides may still be made out. But the battlefield is grown over with grass. One may find evidence of the fighting that once took place, perhaps bullet marks or shell holes. But the fighting ceased long ago. The battle is a thing of the past."

textual traditions referred to and constructed a world. Whereas the old path took a largely epistemological direction and relied heavily on comparison between Babylonian and Western sciences, using either Greek or even modern science as models and templates, the new path points toward the anthropology of science and questions of ontology as well as epistemology. The project is not to replace the old with a new methodology but to complement and augment existing methodologies for the historiography of the cuneiform sciences.

BROADLY CONSTRUING SCIENCE

To speak of the *sciences* of the cuneiform world undoubtedly seems controversial and even presumptuous to many. Before going too far into this introduction, therefore, it would be well to clarify my use of the term science. Of course, no one- or two-sentence definition will ever adequately account for something as complex, multifaceted, and historically rich as science, but for my purposes the term science will refer to cultural activities that establish normative practices and methods of inquiry into, and result in systems of knowledge of, what a particular community imagines and perceives as its phenomenal world. Of central importance for defining science, in the sense I employ it here, is both how knowledge of phenomena is acquired (through observation, classification, modeling, reasoning) as well as how and why certain phenomena become objects of inquiry.

Implicit in this definition are the members of an extended intellectual community of scribes whose perceptions and imagination concerning the phenomena of their interests are reflected in the corpora of texts they termed *ṭupšarrūtu* "the art of the scribe" (Robson 2019). The point of departure for the historian and the anthropologist of the cuneiform sciences is the extensive corpus of written material this community of scribes produced over the long span of the cuneiform tradition. It is worth emphasizing that however we characterize the cuneiform culture of science, it was rooted in texts, written, copied, and preserved by *scribes*. Therefore, no matter what we say about how the cuneiform scribes thought or what they knew, whether in the Old Babylonian, Neo-Assyrian, Neo- or Late Babylonian periods, we are speaking of an intellectual culture attached to educated literate elites, not all or general "Babylonians" or "Assyrians" or "Mesopotamians." To paraphrase Geertz (2010, 18) on Malinowski's Trobrianders, such thought or knowledge is not meant to be generalized or essentialized as something "The Babylonian" or "The Mesopotamian" would think, know, or believe by virtue of being a Babylonian or, much less, a "Mesopotamian." The cuneiform script as the medium for many sciences spanned millennia and various regions and accommodated a variety of scientific traditions.

In addition, unlike in the Classical Greek and Roman world, where individual names are often attached to traditions, especially philosophical and scientific traditions (such as the Pythagorean, the Hippocratic, or Aristotelian), the cuneiform scribes are known to us only as copyists and editors of tablets and therefore tend to be referred to (by us) collectively as "the scribes." During the seventh century, the scribes who produced and used the texts that *ṭupšarrūtu* comprised were court appointees and advisors to the Assyrian monarch in Nineveh. Even they, in citing an authority or authorities in the transmission of knowledge, did not mention scribal authors or names but used the term "scholar/s" or "expert/s" (^{LÚ}*ummânu*/*ummânū* pl.). From their many letters and reports on ominous phenomena to Kings Esarhaddon and Assurbanipal of Assyria (Hunger 1992; Parpola 1993), we know many scribal names, but as to who the authors of scientific texts were by name, we remain largely in the dark.

Following the fall of the Assyrian Empire in 609 BCE, the scholarly scribal culture in the period from the sixth century onward moved south into the major cities of Babylonia, mainly Babylon and Uruk, and into the temples of Marduk/Bēl (Esagil) and Anu (Rēš). Late Babylonian textual sources, however, are not supplemented by an epistolary corpus between king and scholar such as is extant between the Assyrian monarchs and their scholars. For identifying the scribes of the late scientific texts where astronomical text colophons survive, we have names of tablet owners and tablet copyists, expressed as "tablet of X" (*ṭuppi* X) followed by "hand of Y" (*qāt* Y) (Ossendrijver 2011b, 213–15), but these scribal designations are a far cry from authorship.

Individual authorship in cuneiform antiquity is a rarity, generally reserved for gods, kings, or legendary early scribes (Frahm 2011b, 520). Some scribes were more individually celebrated than others, such as the second-millennium exorcist Sin-leqi-unnini, associated with the standard Akkadian Epic of Gilgamesh (George 2010, 6; Frahm 2011b, 517), but the individuals in question in the present study are absorbed largely anonymously into one or another intellectual community of scribes active within the cuneiform culture of learning, scholarship, and the sciences. During the Neo-Assyrian period, the upper echelon of scribes was filled to a considerable extent by members of elite families, such as that of the venerable Nabu-zuqup-kena who worked in the court of Sargon II at

Nimrud. This scribe is known as the copyist of the Twelfth Tablet of the Epic of Gilgamesh (Frahm 2005) and numerous scholarly tablets (Robson and Stevens 2019, 325 tablet 8.2 and 337–8). In spite of our knowledge of the names of some prominent scribes, the question of authorship is still moot, and the attachment of names to texts such as we are accustomed to seeing in the Greek and Greco-Roman worlds is not found.

The definition of science as given above can in principle pertain to a great number of organized bodies of cuneiform knowledge, including the material found in the lexical text tradition (Veldhuis 2014; and see Chapter 5). The lexical corpus compiled and systematized terminology pertaining to a great variety of subjects, not only words for "things" but also grammatical paradigms and legal expressions (e.g., *ana ittišu*, meaning something like "at its proper [or agreed upon] time" or "to take account of its circumstances"; Landsberger 1937; Veenhof 2021). The earliest (late fourth and third millennium) lists of words and expressions reflect the interests of scribes dealing for the most part with administrative and bureaucratic matters, whereas a greater variety of lists and their diverse purposes emerge in the Old Babylonian tradition (Veldhuis 2014, 13–14, and throughout). The lexical lists afford a wealth of words for cultural kinds, or what Nelson Goodman called "relevant kinds" (1978, 10) to distinguish cultural from natural kinds.[2]

Although Sumero-Akkadian lexicography exemplifies systematized knowledge in our sense, it was not part of the project developed by the Babylonian scribes for the investigation of phenomena as ominous signs and for the theorization and mathematical (or nonmathematical) prediction of periodic phenomena. It was not included among the corpora the scribes designated under the rubric *ṭupšarrūtu* (see Chapter 2), which refers to scholarly and technical knowledge. The texts of *ṭupšarrūtu* were systematized and, like lexical texts, display the vertical (paradigmatic) and horizontal (syntagmatic) relationships between the parts (i.e., words, phrases, or clauses) of the lists. Omen lists even depend upon some of the philological methods employed in cuneiform lexicography. However, the scholarly and technical knowledge of *ṭupšarrūtu* extended

[2] Goodman explained as follows: "I say 'relevant' rather than 'natural' for two reasons: first, 'natural' is an inapt term to cover not only biological species but such artificial kinds as musical works, psychological experiments, and types of machinery; and second, 'natural' suggests some absolute categorical or psychological priority, while the kinds in question are rather habitual or traditional or devised for a new purpose" (Goodman 1978, 10). This describes quite well the nature of the "kinds" and classifications of words and things found in cuneiform lexicography.

beyond words to the world of phenomena and as such have a different epistemic profile.

The interests of late second-millennium, seventh-century, and Hellenistic-period scholar-scribes in certain ominous phenomena led to the establishment of compendia on divination, astronomy, astrology, magic, and medicine (Rochberg 2004, 44–97). The social and intellectual contexts for the development of these interests differed from one period to another, but the textual traditions of the compendia carried over from one period and region to the next to produce what can be referred to collectively as the cuneiform sciences. These include celestial and terrestrial omens (including human and animal behavioral omens, malformed births both human and animal, human physiognomy, dreams, everyday occurrences in and around the city and in houses) and omens from the liver and entrails of sacrificed animals, as well as diagnostic medical texts and magical procedures to deal with both omens and disease. Some of the scribal masters, *ummânū* "experts" or *eruditi*, developed mathematical astronomy in the period after 500 BCE. Some were expert in multiple text traditions. Which scribal families and individuals are associated with which body of texts is a matter for the sociological investigation of the practice of these sciences (Ossendrijver 2011a and 2011b; Robson 2019).

In raising the question of what the phenomenal world consisted in for the Babylonian *eruditi*, we encounter, in a way perhaps not dissimilar to the anthropologist of non-Western contemporary cultures, a different conceptual and terminological color palette from our own. The terms for phenomena in the cuneiform sources take us outside the usual framework of the physical phenomena that typically occupy historians of natural science. Regardless, the relationship between the sources for scribal scholarly knowledge in cuneiform and the history of science is an important one and should not be sidelined for lack of an equivalent term in Babylonian Akkadian, the language of the scribes. Indeed, to exclude the cuneiform materials is to put an indefensibly narrow lens on our historiography of science.

The term science, both by etymology (English science from Latin *scientia*, Middle French *science*) and historical usage (from the fourteenth century), has often been interchangeable with knowledge, or *a kind of* knowledge. As the Byzantinist Stavros Lazaris explained (2020, 7),

> Even though the word is of Latin origin, "science" is a modern not an ancient category. It derived from the verb *scire* "to know," and its adjectival present participle form, *sciens*, "who knows, is educated, has knowledge of." The

substantive *scientia*, "knowledge, science [in the broad sense]" is abundantly attested in classical Latin.

On the other hand, in Lazaris's words (2020, 7–8),

> In Greek, the term *episteme* refers to knowledge that is true and certain, related to the verb *epistasthai*, knowledge, whose etymological meaning (*epi* [over] + *histasthai* [to stand]) is "to be placed above," that is to say, "dominate" the subject of study . . . Besides *episteme* (knowledge) in ancient Greek, one also finds the terms *philosophia* (love of wisdom, philosophy), *theoria* (contemplation, speculation) and *peri physeos historia* (a survey of nature). They are all employed in specific contexts where their translation by the word "science" is (or at least should be) obvious and hardly risks misleading the reader. But if these terms are used to designate certain intellectual disciplines we think of as scientific, each time the term is used it may mean something quite different from our term "science."

The terminological subtleties within the Byzantine epistemic context are relevant to our approach to cuneiform studies, not because they are the same, but because "knowledge" or "science" understood on the basis of Greek words and their root derivatives in European languages takes on even clearer distinctions in the cuneiform epistemic context where the language is Akkadian. The philological and semantic particulars have been discussed before (see Rochberg 2016a, 61–69), so need not be repeated here. What should be repeated is that *ṭupšarrūtu* in its own context of meaning and usage is the best guide to the sciences in the epistemological landscape of cuneiform culture.[3]

The kind of knowledge associated with science was for much of modern history and philosophy of science defined in terms of methods and goals exemplified by early modern science (McMullin 1984a) and consequently was never well suited to categorizing the Babylonian culture of science and its particular epistemic values. With those methods and goals in mind, Ernan McMullin's summary of the developmental course of science classified Babylonian astronomy as predictive in contrast to Greek astronomy, which was both demonstrative and aimed at causal explanation. He put each under the rubric of natural science, and while much of his discussion of both Babylonian and Greek science focused on

[3] The term "cuneiform culture" was first utilized by Karen Radner and Eleanor Robson (2011, xxvii) to refer to a conceptual framework constructed by cuneiform texts and cuneiform writing technologies "fundamental to a modern academic understanding of the region [of the ancient Middle East] but which also bound the ancient inhabitants into a shared set of ways of understanding and managing their world." The cultures of science, knowledge, and scholarship that I focus on in these pages represent closely intertwined dimensions of the broader construct.

astronomy, he made mention of medicine as further evidence of the interest of the Greeks in causes, quoting Geoffrey Lloyd on *The Sacred Disease* (McMullin 1984a, 41 quoting Lloyd 1979, 16). In dealing with comparative bodies of evidence, Lloyd has long favored a broad conception of science, as he said, "closer to the French '*science*' or German '*Wissenschaft*'" (2012, 3), which terms still emphasize the epistemological over other aims of science. What I find useful in his conceptualization, however, is the willingness to consider as science the investigation of phenomena for more diverse reasons and interests and by more diverse methodologies (most recently in Lloyd 2021a) than those found in the modern era.

McMullin's sharply contrastive schema dividing Babylonian predictive science (his so-called P-science) from Greek demonstrative science (so-called D-science) was devised a few years too early to take into account the discovery within Greek astronomical papyri of a whole set of predictive methods for producing tables and horoscopes based on Babylonian systems of linear arithmetic calculation (Jones 1999). Whereas we had long known that Greek cinematic astronomy made use of Babylonian parameters and observations, the internal complexity of the Greek astronomical tradition, in which the cinematic and geometric hypotheses of the planets coexisted with and indeed preserved Babylonian arithmetic methods, was only understood from 1988 (Neugebauer 1988; full treatment in Jones 1999). From the standpoint of methods, the comparative picture of Babylon and Greece turned out not to be so stark nor evolutionary in nature from less to more "accurate." The evidence did not support either an evolution from predictive to demonstrative sciences or a definition of science that depended upon such a characterization.

Raymond Williams also discussed the many meanings of the term science and its evolution over time (Williams [1976] 1983, 276–80). As McMullin would later do, Williams began his definition methodologically and emphasized first demonstration and then, as a later development, empiricism and experiment. He said: "Changes in ideas of Nature encouraged the further specialization of ideas of method and demonstration towards the 'external world,' and the conditions for the emergence of science as the theoretical and methodical study of *nature* were then complete" (Williams [1976] 1983, 278). This way of understanding the meaning of science was well hardened by the early to mid nineteenth century, having begun already a century before, and scientific knowledge meant the result of inquiry into physical phenomena alone, expressly excluding theological and metaphysical inquiry. The equation of the "external world" with nature is a feature of this understanding of science.

The scope and central goals of science as being one with naturalism – where "the external world" is one and the same as nature, repudiating and eliminating supernatural phenomena as not belonging to the external, or real, world – according to Peter Harrison (2019, 6–7) originated in the European Middle Ages, although with vastly different causal implications. As Harrison explained (2019, 8):

> Thomas Aquinas (1225–74) was to popularize the term 'supernatural' (*supernaturalis*) to label this mode of divine action [i.e., outside the bounds of the causes God created in nature]. The natural–supernatural distinction thus began to crystallize in the thirteenth century as a means of distinguishing two kinds of divine activity: one in which God works with the order he embedded into things; the other when he acts miraculously and independently of created causes.

The distinction between kinds of divine causality has little to do with modern naturalism, where reality and nature become one, without legitimate ontological space for immaterial, "supernatural," things. This would not become the hardened view until the nineteenth century. In the Renaissance, we still encounter a remark such as that of Ficino, citing Plotinus and Synesius: "Nature is a magician"; and natural magic and astrology, together with the reality of stellar agency and cosmic sympathy, were parts of natural philosophy (Copenhaver 1988, 274).

A sharp contrast between naturalism and the epistemological manifold[4] reflected in cuneiform scholarly texts has been delineated and explicated elsewhere (van de Mieroop 2016; Rochberg 2016a, 2021). In spite of the acute contrasts and discontinuities with the natural philosophy of Greek, Greco-Roman, and later European traditions, there are also similarities due to the extensive intellectual heritage from the cuneiform Middle East in those (as well as other) traditions.[5]

[4] My use of the term "epistemological manifold" in the context of *ṭupšarrūtu* is inspired by, but differs from, the cultural manifolds of Geoffrey Lloyd and Nathan Sivin's comparative work on ancient Greece and China (Lloyd and Sivin 2003; Sivin 2005). These refer to the conceptual frameworks, epistemic norms, and social institutions that supported science in the cultures of ancient Greece and China as well as the entire complex that unified these parts in their different contexts. What I mean by epistemological manifold also makes a whole out of various parts but is in reference to the various knowledge corpora and methods of knowing constituted by the multifarious disciplines of *ṭupšarrūtu* and its changes over time.

[5] As far as other ancient intellectual traditions of astronomical, astrological, and divinatory content are concerned, viz., corpora preserved in Egyptian, Aramaic, and Sanskrit, the contribution of the cuneiform sciences to those traditions of divination, astronomy,

In view of the diverse relationships of likeness and difference, continuity and discontinuity across the sources for the history of science, science cannot be reduced, unified, or universalized from its many historical contexts and communities in order to serve a definition such as that of Williams quoted previously. Regarding the kind of scientific inquiry attested in cuneiform texts, to reduce, unify, and universalize that evidence such that it instantiates such a definition of science results in an erasure of its actual nature and character for the sake of its resemblance to a few aspects of science so construed. Knowing nature, and establishing an equivalence between the external world and nature – the hallmark of naturalism – is only one of multiple ways science has provided a means for inquiry into all kinds of phenomena over the course of its history. In ancient Babylonia, science had other aims.

The problem of classifying cuneiform texts such as omens and medical astrology as science is sometimes addressed by replacing the word science with the word knowledge as less loaded and not as anachronistic. I find the word knowledge to be a poor substitute in this case, as it is too general to convey certain dimensions of the cuneiform scientific corpus that are in fact continuous in a variety of ways with later sciences. On that basis I defend the use of the term science and find it has heuristic value. In the case of ancient Babylonia, a more integrated approach where questions of epistemology are counterbalanced by those of ontology requires the all-important translation of the Babylonian conceptual framework (how and what phenomena were identified and understood) and practical framework (that is, what activities produced knowledge, such as observation, model-making, and various forms of schematization and prediction), and thus the translation of Babylonian epistemic and ontological norms (Severi 2015, 76–81) to ones we can understand. The writers of cuneiform scientific texts did not think they were practicing "science" (far from it). But with Nicholas Jardine (2000, 252), as well as Jan Golinski (2012, 36) and Peter Dear (2012, 50), I find legitimate the historical use of our modern term science for the cuneiform compendia and their implied practices within a bigger picture of the history of science overall.

Furthermore, the use of the term science does not sanction modern European standards for defining what scholars can or should investigate under the rubric science. For historians of premodern scientific traditions,

astrology, and medicine is well known. For the relationship of the Babylonian material, in the form of a Greco-Babylonian influence, further to the east, in astronomical and astrological texts from China as well as Japan, see Mak 2019, especially pp. 25–34.

this critical stance on the term science was already articulated by David Pingree in his now classic 1992 paper "Hellenophilia v. History of Science." Among the several false propositions he countered in this paper, I cite here only two: first, that "the only real sciences are those that began in Greece" and, second, that "the true definition of science is just that which scientists happen to be doing now, following a method or methods adumbrated by the Greeks, but never fully understood or utilized by them." Pingree chalked this up to "cultural myopia," which also goes by the term "Eurocentrism" or "ethnocentrism" (1992a, 555; and see Chapter 5).

If we are to move beyond an exclusive focus on epistemology and begin to inquire into the ontological framework for cuneiform knowledge, then it makes sense to look to the second of the twofold thematic of this book, namely, the anthropology of science. If the way in which I have carved out the scope of an anthropology of historical sciences for this project is justified, namely that consideration is given to the relationship a culture strikes with its own world, then the question "To what sort of world did the cuneiform sciences refer?" is an anthropological question.

The next section sets out the background and foundation for my application of the anthropology of science to the historical situation of ancient Babylonia.

THE ANTHROPOLOGY OF SCIENCE

Bruno Latour (1990) set terms for an anthropology of science aimed not at "the Tropics" but rather at modern Western science, arguably an early form of what came to be called "the anthropology of the near" (Augé 2006). This was the culmination of Latour's landmark two-year study of the scientific culture of the Salk Institute for Biological Studies, which he wrote up together with sociologist Steve Woolgar (Latour and Woolgar [1979] 1986). In this ethnographic excursion to a "tribe of scientists," the term "anthropologist of science" first appeared, as well as "anthropology of science," each in quotation marks (Latour and Woolgar [1979] 1986, 17, 19, and 27, respectively).

Latour distanced himself from the expected manner in which ethnographic or cultural anthropology had an interest in science, saying, "Since the time of Lévy-Bruhl, anthropology has always been interested in science, but in the sciences of the Others ... putting to use the methods of anthropology in order to understand our sciences is only recent" (1990, 145). Latour cited Jack Goody's *The Domestication of the Savage Mind* as marking "the beginning of the anthropology of science," in which the Lévi-Straussian dichotomy of hot and cold societies would be dismantled in favor of other kinds of divides – "writing, practices of list making, skills at handling proto-libraries" – but still aimed at nonmoderns (Latour 1990, 146). Nonmoderns were the object of inquiry, but nonmoderns in the contemporary world.

The present project may well be a return to the interest in "the sciences of the Others," but with a perspective firmly rooted in the idea of science as a cultural phenomenon and a new purpose for "old" questions to inform our historiography of premodern sciences. By "science as a cultural phenomenon," I mean that science, its methods, and its results, indicate directly how communities find order and intelligibility in the phenomenal world of their experience and imagination.

There is an implicit plurality to the statement just made, namely of a plurality of phenomenal worlds, that is to say conceptions of phenomenal worlds tied to the historical communities who conceived and organized them. The term phenomenal world was used by Paul Hoyningen-Huene (1993, 31–41, also Sankey 2000, 64 and throughout) in discussing Thomas Kuhn's *Structure of Scientific Revolutions* ([1962] 1996). Hoyningen-Huene limned out what he called "the world concept" in Kuhn's text as falling into two distinct constructs, one as the immutable world-in-itself and the other as a variable phenomenal world, variable within paradigms. In Kuhn's usage, both world concepts assumed the identity of the world and nature. The phenomenal world was a perceived world within a paradigm, whereas the other sense of the word 'world' was as the fixed object of scientific inquiry, not subject to change by revolutions in science, in spite of its being seen "in a different way" in the aftermath of such revolutions (Kuhn [1962] 1996, 53). Further discussion of Kuhn's world/worlds will be of interest in Part II, Chapters 4 and 5. It is worth emphasizing here, however, that when we turn to the sciences as preserved in texts of the ancient Middle East, the identity of world and nature will not serve our analysis for lack of the conception of nature in the cuneiform sources for science.

In addition to the problem of plural worlds for the history of science after Kuhn, namely in what sense they are both epistemologically and ontologically many, or indeed whether plurality applies only to the world's many representations, pluriworldism became the subject of philosophical discussion (McCormick 1996, with previous literature) centered on Nelson Goodman's 1978 *Ways of Worldmaking*. A more recent but no less intense anthropological discussion concerning multiple worldviews (a multiplicity of conceptual schemes reducible to a single underlying reality) versus pluriworldism (a multiplicity of irreducible realities) percolated through the last decade and continues until today (Venkatesan 2010; Alberti et al. 2011; Heywood 2012; Paleček and Risjord 2012; Pina-Cabral 2014a and 2014b; Viveiros de Castro 2015; Jardine 2019; Tola and dos Santos 2020; Tola and Medrano 2020; Lloyd and Vilaça 2020; McCarty et al. [2021] 2022, Lloyd and Vilaça 2023).

Although the anthropologists do not often cite Goodman, his 1978 *Ways of Worldmaking* traced the question of many worlds to the philosophical anthropology of Ernst Cassirer's multiplicity of worlds reflected in comparative symbol systems.[6] Cassirer's plurality of worlds was no mere possible worlds but actual, real worlds for their inhabitants. Goodman's *Ways of Worldmaking* together with its philosophical foundation is still apposite and, I would say, relevant for anthropologists interested in these questions. Goodman stated the basic premise this way:

> Since the fact that there are many different world-versions is hardly debatable, and the question how many if any worlds-in-themselves there are is virtually empty, in what non-trivial sense are there, as Cassirer and like-minded pluralists insist, many worlds? Just this, I think: that many different world-versions are of independent interest and importance, without any requirement or presumption of reducibility to a single base. The pluralist, far from being anti-scientific, accepts the sciences at full value. His typical adversary is the monopolistic materialist or physicalist who maintains that one system, physics, is preeminent and all-inclusive, such that every other version must eventually be reduced to it or rejected as false or meaningless. (1978, 4)

Therefore, the way the world is thought to be made (what exists) and to function (how it works), which we see in the ways cultures describe their worlds and analyze the behavior of the phenomena of interest to them, is the way of "worldmaking." By worldmaking, Goodman did not mean a cognitive creation *ex nihilo* on the part of each culture but rather a rethinking, reconceiving, and reorganizing on the basis of old ways of thinking, conceiving, and organizing. In his words, "The making is a remaking" (1978, 6), and "Worldmaking begins with one version and ends with another" (1978, 97).

[6] Regarding Goodman's relation to Cassirer, Paul Ricoeur, in his review of *Ways of Worldmaking* (Ricoeur (1980, 115) said, "I have no hesitation to acknowledge that I heartily approve the daring attempt to go further than Cassirer in the recognition of the plurality and irreducibility of world-versions. This thesis is not only liberal but liberating. To go further than Cassirer, Goodman had to give up the still hierarchical conception of symbolic forms whose outcome was to put the scientific version at the top of the ascending scale of forms."

The relationship between world and worldmaking ranks as one of the fundamentals of anthropological interest and is not limited to modern, or Western, or non-Western contemporary contexts but can be extended to historical sources for sciences in non-Western and premodern contexts. Therefore, in addition to the anthropology of modern science, which was Latour's project on how science is produced in contemporary industrial and laboratory contexts, there is a relatively unexplored anthropology of science that extends to historical contexts. By the same token, inquiry into the relationship between world and worldmaking also has historical pertinence and productive implications for historiography.

How to define and theorize worldmaking, therefore, is a crucial ingredient in the present historiographical project. Following Goodman, although he did not have a historical agenda, worldmaking raises the question of the ontological nature of historical worlds and offers a way to legitimize the ontologies of such worlds as investigated by the sciences of their day. Otherwise, we are forced to view the worlds referred to in ancient texts not as worlds-in-themselves but as faulty and imperfect worldviews, misbegotten perspectives on and representations of the one (true, real) world of physical science. Accordingly, the sciences of non-Western and premodern cultures can only be seen as wrong and their ideas about phenomena either as missing the mark or as absurd fabrication. This seems to me to be a rather unforgiving historiographical attitude.

The ethnographic *in situ* aspect of Latour's anthropology of science is obviously impossible for cuneiform culture, but the exploration of worldmaking in cuneiform sources belongs more to an anthropology of science than it does either to sociology or philosophy of science. In the case of Babylonia (and Assyria), the participants in question are of course not "scientists" but scholar-scribes, professionals with titles such as *ṭupšar Enūma Anu Enlil*, "scribe of the series titled *Enūma Anu Enlil* (for celestial divination)," or *kalû*, "incantation priest," and the contexts are not laboratories or universities but the royal palace court at Nineveh and Kalhu or the temples of Marduk in Babylon (Esagil) or of Anu in Uruk (Rēš). The object of inquiry for the anthropology of science in the cuneiform context is not limited to the social aspect of knowledge production nor merely the fact of bodies of systematized knowledge. Rather, it is the describing, imagining, and representing of the experienced and known world in the texts produced by the cuneiform scholar-scribes that become the focus of analysis (see Chapters 5, 6, and 7).

ONTOLOGY AND THE NOTION OF WORLDS

I have referred to ontology and the notion of worlds as significant to the historiographical project of this book. It is time to define in all brevity what I mean (further development will be found in Part II). In twentieth-century Anglo-American philosophy of science, Willard von Orman Quine viewed ontology in terms of the referents, the objects posited by scientific theory, saying in the 1948 paper titled "On What There Is":

> A curious thing about the ontological problem is its simplicity. It can be put in three Anglo-Saxon monosyllables: "What is there?" It can be answered, moreover, in a word – "Everything" – and everyone will accept this answer as true. However, this is merely to say that there is what there is. There remains room for disagreement over cases; and so the issue has stayed alive down the centuries. (1948, 21)

This statement has more the quality of a Zen kōan than it does of analytic philosophy, but historians and philosophers of science and anthropologists alike have taken it as a point of departure (e.g., Heywood 2012). Reframing Quine's statement that "there remains room for disagreement over cases" – which concerns the underdetermination of scientific theory by empirical evidence – to anthropological concerns, indeed, once one gets down to the "cases," we can be deep in cultural difference.

The cultural differences between "cases" are the remit of anthropology, and the field readily took up the ontological challenge. Amiria Henare, Martin Holbraad, and Sari Wastell, in *Thinking through Things* (2007), first used the phrase "ontological turn," calling it "a quiet revolution" (Henare et al. 2007, 7) that took root in anthropology during the 1980s and 1990s associated with Bruno Latour, Alfred Gell, Marilyn Strathern, Eduardo Viveiros de Castro, and Roy Wagner. They challenged the idea of anthropology as "the *episteme* of others' *epistemes*" (Henare et al. 2007, 9) and began to show how ethnographic evidence could unsettle conventional anthropological ideas about worldviews and representations, and in so doing converted epistemological to ontological questions. Lily Chumley noted that "the ontological turn, which began with ethnographies of Amazonian worlds, links to an environmentalist concern to theorize 'other kinds of realities beyond us' (Kohn 2014), an *us* that might refer to modern, Western, or otherwise privileged subjectivities, to the human species, or to sentient life forms generally" (Chumley 2017, S1).[7]

[7] That those *other kinds of realities beyond us*, especially in the literature on materiality, focused attention on animals and other nonhuman entities (Chumley 2017, S2) "beyond

The tradition of the anthropologists who forged and developed the ontological turn (Wagner [1972] 1981, Strathern 1980, 2004, Descola [1986] 1994, Latour [1991] 1993, 2004, Viveiros de Castro 1992, Ingold 2000, Holbraad and Pedersen 2017) is not, as Tola and dos Santos observed (2020, 170), homogeneous. Some (famously Viveiros de Castro) find multiple ontologies in multiple different real worlds, while others (Holbraad and Pedersen) see the literal reading of multiple ontologies and many worlds as "preposterously reifying" (Holbraad and Pederson 2017, 176), preferring to take ontology as a *singulare tantum* noun. Holbraad and Pedersen's project is methodological and refers to how anthropologists ask their questions, namely, "to ask ontological questions without taking ontology as an answer" (Holbraad and Pedersen 2017, 11). They not only oppose "taking ontology as an answer," they view the descriptors for Indigenous ontologies or cosmologies, such as the Melanesian, Amazonian, or Mongolian, as "territorializing and essentializing," and the idea of many worlds as "of no help" (Holbraad and Pedersen 2017, 176).[8]

With regard to "territorializing and essentializing" descriptors, my use of the term cuneiform science is in part meant to obviate the problem. But will an anthropology of science as a methodology for considering the ontology of the phenomenal world implied in cuneiform scientific texts entail the kind of essentialist reification called out by Holbraad and Pedersen? Is it enough to disavow the use of the terms "Babylonian" or "Assyrian" as essentializing and then proceed to examine ontologies in "the cuneiform world" from texts in the cuneiform script produced by a community of scribe-scholars (Assyrian and Babylonian astronomers, diviners and *kalû* priests) without falling into essentialist reification?

and below human capacities of representation" is clearly a different project than that of historical sciences.

[8] This critique impugns the study of non-Western people and their particular "realities" in the sense, as articulated by Lewis and Wigen (1997, 114–15), that "in presuming to speak for and about foreign peoples, the Western observer can cast them only into the role of an objectified, voiceless 'Other.' The triumph of such a scientistic (and masculine) mode of seeing (or "gaze") – i.e., the continuing value placed on knowledge gained though invasive scrutiny by expert outsiders – is widely denounced today as a pernicious legacy of imperialism." Lewis and Wigen point out, however, that while this sustains Eurocentrism, its more troubling implications are, and here they quote Samir Amin (1989, 147), that "only Europeans can understand Europe, Chinese China, Christians Christianity, and Moslems Islam." This, they say, is "to surrender to an ethnocentric world" (Lewis and Wigen 1997, 115). See also below, Chapter 5.

Perhaps further questions can also be raised. Is it territorializing or essentializing to speak of an ancient Greek and medieval natural philosophers' model of the world as a finite unitary set of nested geocentric spheres comprised of the elements and the celestial aether? Not all Greek and medieval natural philosophers modeled the world or identified its ontology in that way, and ideas changed over the fifteen centuries from Greek antiquity to the end of the Middle Ages. But in the main, the Ptolemaic cosmos persisted for 1,500 years in the West, surely a territorial consideration for historians. It has turned out to be a mistake to write ontology out of that model, as Bernard Goldstein showed (1967; 1980, 136; and see Chapter 7). With his edition of Ptolemy's *Planetary Hypotheses* (1967), Goldstein was able to restore the ontological reality of the arrangement of and distances between the Ptolemaic planetary spheres and the relative sizes of the planets in the Ptolemaic cosmos. In that instance, ontology – as an answer – was always a useful category of analysis for a scientific world model, providing an answer to the question of what exists vis-à-vis the heavenly spheres in ancient Greek and medieval cosmologies. Apart from ancient Greece and medieval Europe, when it comes to matters of the history of science, especially of the remote and distant sciences, and specifically of cuneiform sciences, ontology remains underutilized in our analyses of historical texts and the historical motivations for inquiry as reflected in those sources; when it *is* utilized, it is the ontology of nature and of natural phenomena that is "of no help."

Grounding and fueling Holbraad and Pedersen's criticism of the anthropology of ethnographic ontologies as essentializing and reifying is their alliance with the posthumanist twenty-first-century development of object-oriented ontology (Holbraad and Pedersen 2017, 33–36), according to which the existence of objects is not dependent upon human frameworks of knowledge or human interaction. What Holbraad and Pedersen's alliance with object-oriented ontology means for their conception of "world" is the shared goal to decenter the human in order to posit a realm of existence for objects independent of the human mind to conceive them or human activity to relate to them (Holbraad and Pedersen 2017, 35). Even taking into account ways in which object-oriented ontology differentiates itself from traditional metaphysical realism, its claim of mind-independence for objects lands it squarely within metaphysical realist territory. Mind-independence, and the realism it underpins, has been rejected by pragmatists and pluralists of a number of stripes, Hilary Putnam being one (see Chapter 4, note 10), more recently Hasok Chang (2022, 68–83).[9] Goodman's philosophy of world-making and his argument for "many worlds if any" stands as perhaps the most forceful rejection of mind-independence. Chapter 4 develops this line of my argument and provides philosophical context for taking steps toward an anthropology of early sciences.

The lack of homogeneity about the ontological turn among anthropologists and the fact that a consensus has still not been reached concerning the referentiality, the *reality* of other ontological systems and the worlds that answer to them, makes summarizing some of the main voices in the debate worthwhile. To begin, the question of multiple ontologies was taken up as a motion tabled at the Group for Debates in Anthropological Theory at the University of Manchester in 2008 before an audience of sixty-four (Venkatesan 2010, 199). The discussion appeared in 2010 in *Critique of Anthropology* 30, where the motion read: "Ontology is just another word for culture." Two sides of the debate were initially posed such that one side saw many and divergent cultural representations of a single reality; the other, dubbed the ontological approach, saw many ways of worldmaking justify the idea of multiple realities and a plurality of worlds (Candea 2010, 174). Although all were concerned with how to define more precisely the relationship between culture and ontology, whether they do similar or different things, few discussed pluriworldism directly. Neither realism, idealism, object-oriented ontology, nor any particular philosophical orientation figured in the debate. In the end, the motion lost to an overwhelming majority against it. Most did not like the connotations of the word "just," finding that "ontology is another word for culture" (without "just") had possibilities.

The following year saw the publication of another forum in anthropological theory (Alberti et al. 2011) addressing the question of ontological difference from an archaeological perspective. The published forum, consisting of five responses from a 2010 session "Worlds Otherwise" of the "Theoretical Archaeology Group," began with the acknowledgment of the "dramatic increase in academic labor concerned with ontology" and provided a long bibliographical list of works (Alberti et al. 2011, 896). The turn to questions of ontology, they said, had to do with the nature of reality and being, "provoking an upheaval in how various disciplines have conceived of agency, change, causality, materiality, and relations" (Alberti et al. 2011, 896) and constituting "an immense project of metaphysical renegotiation"

[9] For a counterposition, see McMullin 1984b.

(Witmore in Alberti et al. 2011, 897). Because archaeological source material is located in historical contexts and historical "things" (tombs, temples, walls, sarcophagi, pottery, and so on), the archaeologists' answers to the ontological question – some for and some against – are an important addition to the otherwise philosophical and ethnographical discourse on this subject. The archaeologists' ideas are perhaps more relatable for cuneiformists, who also deal in historical things, but via the references to and descriptions of those things in texts. I note here, from Fowles's remarks, that "every discipline is fundamentally engaged in the study of things, be they textual things, psychological things, stone and bone things, or what have you. (To say otherwise is to play into the very boundary between things and concepts that many seem so invested in overcoming.)" (Fowles in Alberti et al. 2011, 899). It is in relation to the engagement with the ontology (both of being *and* becoming) of things that Marshall's argument (in Alberti et al. 2011, 900) for why archaeology is anthropology resonates with the position I adopt here. Namely, that by virtue of our study of the "things" of Babylonian scientific culture (stars, eclipses, acronychal risings, symptoms of disease, magical preparations, demons and other divine causal agents), cuneiform studies in the history of science also has an anthropological dimension in which the option of engaging with an other ontology and an other world presents itself.

A parallel engagement with multiple ontologies also emerged in science and technology studies (STS), where it was discussed in a special issue of the journal *Social Studies of Science* 43/3 titled *A Turn to Ontology in Science and Technology Studies?* edited by Steve Woolgar and Javier Lezaun (2012–13), and then in *Social Studies of Science* 45/3 (2015) (Sismondo 2015, Aspers 2015, and Woolgar and Lezaun 2015). Sergio Sismondo noted that "to the extent that it develops ontology as an explicit study, STS does so via insights about multiple ontologies" (Sismondo 2015, 443). In 2017, Andrew Pickering returned to the subject to detail his "own attempt to take different worlds seriously" (Pickering 2017, 135) and showed that the reckoning with ontologies in STS differed from that of the anthropological debate in that pluriworldism was more directly confronted.

The lineage for the ontological approach in STS and the history of science can be traced to the first formulation of the social construction of reality in the 1980s and then in the widespread use of the construction metaphor in the history of science. Historical ontology, in Ian Hacking's monograph of the same name (2002), had to do, among other things, with phenomena being created, ontology as contingent on human engagement with things, making things/facts – "The factization of the world is a human activity," Hacking said (2002, 65). Constructivism (or social construction), as Sismondo noted (2015, 442–43), afforded a view of ontology for STS not as a fixed order of things standing immutably outside human endeavor but as contingently derived from scientific practice. Pickering also pointed to the constructivist lineage for STS's reckoning with multiple ontological worlds, but placed the ontological "turn" in STS in the twenty-first century after the social constructivist consensus broke down (Pickering 2017, 135). It was at that point, according to his reconstruction, that pluriworldism, the serious consideration of the possibility of actual different real worlds, was on the table.

After all these long-standing debates over ontological relativism and multiplicity within philosophy, anthropology, and science studies, the ontological dimension of non-Western "other" sciences, such as those of the cuneiform record, provides a new and potentially fruitful direction for the historiography of early science. Repudiation of the history of science as the single story of humanity's pursuit of knowledge of the natural world, the world as it *is*, is prerequisite. As Goodman said (quoted p. 14), "the pluralist, far from being anti-scientific, accepts the sciences at full value." Furthermore, he continued, the "monopolistic materialist or physicalist who maintains that one system, physics, is preeminent and all-inclusive, such that every other version must eventually be reduced to it or rejected as false or meaningless" (Goodman 1978, 4) must be superseded. The same idea was later echoed in Eduardo Viveiros de Castro's essay "And," where he said that the supposed epistemological democracy of anthropology was really "based on an absolute ontological monarchy, where the referential unity of nature is imposed" (2003, 18). This has direct relevance for the way forward for the cuneiform sciences.

Each of these articulations resonates with my own conviction that the cuneiform sciences were not in pursuit of natural knowledge and that, consequently, the world to which the cuneiform scribes referred was ontologically "other" in their conception than that which is framed by naturalism and physics. In other words, the Babylonian system was not simply a provincial subject of a reigning absolute ontological monarchy, a regime established by historians upon evidence wholly outside of the cuneiform corpus. As a consequence, the history of science cannot be one long story of humanity's coming to terms with the world as it putatively is according to Western physics, unless we are only interested in how that particular – our particular – form of science evolved from the point when matter and motion became central concerns.

In sympathy with the anthropology of the ontological turn, that is, of those with an interest in other ontologies and multiple worlds, I find untenable the position that nonmodern, premodern, or non-Western knowledge systems merely offer different representations of the one culture-independent, language-independent, and mind-independent, and thus version-independent, real world. When it comes to the cultural form we call science, this seemingly intuitive "truth" robs every other system of world construction besides that of modern science of any ontological legitimacy, a point of view that cannot sustain either anthropology or the history of "distant sciences," as Jardine (2019) called them.

In addition to the implications of multiple ontologies and the claims of pluriworldism for the historiography of science, the subject matter of the anthropological debate intersects at many points with Goodman's worldmaking project, with one significant omission, namely, Goodman's idea of irrealism, which I see as the crux of his claim to the many ways the world can be made, the many ways the world can "be." Certainly no anthropological discussion that I was able to find makes mention of it. Irrealism does not claim that nothing is real or that anything is unreal. It is not a denial of ontology, but an assertion that there is nothing neutral about ontology; "what is" is a function of and dependent upon a world-version.

Goodman's irrealism stemmed from his dismissal of just about every philosophical doctrine, not only realism and idealism but also empiricism and rationalism, preferring "irrealism," *not*, as he said, "as one more doctrine … but an attitude of unconcern with most issues between such doctrines" (1996a, 146). He remarked, "the realist will resist the conclusion that there is no world; the idealist will resist the conclusion that all conflicting versions describe different worlds. As for me, I find these views equally delightful and equally deplorable – for, after all, the difference between them is purely conventional!" (1978, 119). By "no world," he meant the evanescence of worlds and versions, that there is no one way the world is but many ways-the-world-is (Goodman 1978, 4; Leeten 2012). Irrealism not only provides a persuasive argument for pluriworldism and multiple ontologies but, in my view, is also well suited to historiographical purpose.

Goodman's claim about the making of worlds as the making of versions has broad implications for a history of science interested in non-Western premodern cultures such as the Babylonian. He said that even such things as "how the earth moves, whether a world is composed of particles or waves of phenomena, are matters determined not by passive observation but by painstaking fabrication … species do not come prepackaged, separated by unbridgeable gaps or inexorable dictates of nature, but depend upon the relative weight we put upon certain features for given scientific purposes" (Goodman 1984, 21). The physical world as it has been constructed by natural science is, in Goodman's sense, but one world among many. The implications, therefore, for the history of science (and Assyriology) are that it is up to the historian (or Assyriologist) to reconstruct the way of worldmaking for a particular time and place based on the sources describing the things that exist in their ordered world.

If irrealism provides an interesting philosophical basis for historiographical purposes by focusing on descriptions of phenomena rather than on descriptionless preexistent stuff, it has not been embraced by many, certainly not by most philosophers. Indeed, few have addressed it head on. One of the few[10] was Harvey Siegel (1984), who accepted Goodman's making and fabricating as a function of describing what we describe but rejected what he took to be Goodman's further claim denying the very existence of objects of scientific (or any other kind of) knowledge until a version comes along to make them exist. Siegel objected: "It does not follow that the universe is non-existent or devoid of features until we create its existence or its features by our descriptions" (Siegel 1984, 20). What irrealism denies, however, is not ontology as such, but the existence of a discrete fixed order of things (the "inexorable dictates of nature") lying outside (where?) our conceptualizations. Existence *sans* description, in Siegel's formulation, becomes useless from Goodman's irrealist perspective. Irrealism denies a raw *given* in favor of how things are given to us through experience, perception, and reason, that is, what and how things are given (Goodman 1960, 50–51). "Any raw stuff," Goodman said (1984, 34), "is as much the creature of a version as is what is made out of that stuff." Siegel concluded (1984, 20) that, "echoing Hanson,[11] one might say that for Goodman facts are 'version-laden' in an extreme way: strip away all the features of the version, and nothing 'purely factual' remains." But the identification of "factual" objects in a world *is* dependent upon criteria that operate within a version; the identity of facts, features, and objects of worlds are not version-*in*dependent, unless one subscribes to some form of realism.

The point is that outside the contexts in which the classification of a thing is an answer to the question "What is it?" we find ontological

[10] Another was Israel Scheffler, discussed below in Chapter 4.

[11] Referring to Norwood Russell Hanson, who in *Patterns of Discovery* (1958, 2, 19) called the act of seeing "theory-laden," by which he meant "observation of x is shaped by prior knowledge of x" (1958, 20).

imaginaries responding to other questions, such as "How does it work?" "What do we do with it?" or "How has God, or the gods, determined a thing or a being's place in the order of things?" These are relevant cultural questions for the ways human beings organize, navigate, and interpret the things of which their worlds are made, material and otherwise. The importance of considering ontology and pluriworldism is thus a part of the intellectual decolonizing of the sciences of "the Others." Viveiros de Castro defended the turn to ontology as a way to reattach to reality the ontological commitments of non-Western (I would add premodern) cultures. He said:

> I think the language of ontology is important for one specific and, let's say, tactical reason. It acts as a counter-measure to a derealizing trick frequently played against the native's thinking, which turns this thought into a kind of sustained phantasy, by reducing it to the dimensions of a form of knowledge or representation, that is to an "epistemology" or a "worldview." (2003, 17–18)

It seems to me that the philosophical argument to bolster the anthropologists' ontological turn (excepting Holbraad and Pedersen's reworking as methodology only) is Goodman's irrealism. It is the very version-ladenness of phenomena that reinforces the argument against cultures as merely engaging in their own representations of an underlying universal reality. Irrealism's fundamental claim is that there is no "pre-packaged" world, but rather many worlds dependent upon our manner of engagement with them. In this way, irrealism justifies the claim to multiple ways of worldmaking and multiple ontologies. Further discussion of the idea of worlds and its importance for the history of the non-Western premodern sciences is found in Chapter 4.

SUBJECTS AND OBJECTIVES

Part I in three chapters focuses on the historiography of cuneiform knowledge and science. Chapter 1 sets the stage by reviewing aspects of the intersection of Assyriology and the history of science, from the decipherment of cuneiform astronomical texts in 1881 until 1935, when Otto Neugebauer and B. L. van der Waerden entered the field and initiated a second phase. The beginning of the third and most recent phase may be said to have started in the 1990s, when an understanding of cuneiform sciences as including not only mathematics and astronomy but also omen divination and horoscopes came to the fore (Rochberg-Halton 1991; Rochberg 1992; Pingree 1992a). Social, cultural, and epistemological questions began to be asked of the astronomical material with the goal of reconstructing not only the mathematical methods of Babylonian astronomy but its nature as a cultural phenomenon and, as well, the role, status, and place of its scribes in cuneiform intellectual history (Ossendrijver 2011a, 2011b; Robson 2019). This chapter also describes the cuneiform scribal-scholarly knowledge termed *ṭupšarrūtu* in Akkadian as a basis for further discussion of both the culture of cuneiform science and its relevance for reflecting on worldmaking.

Chapter 2 is concerned with some of the chief trends in the history of science that established cultural and pluralist approaches to science essential to integrating the cuneiform scientific textual culture within the history of science and to moving it off the margins and into the history of premodern science proper.

Chapter 3 turns to parse and analyze the historiographical contribution of three prominent historians of science in the mid twentieth century who shaped the modern historiography of the astronomical sciences. They are Otto Neugebauer, arguably the preeminent contributor to our knowledge of Babylonian astronomy; Olaf Pedersen; and David Pingree, whose work on Greek, Latin, Sanskrit, Arabic, and Akkadian astrological texts brought astrology within the domain of the history of science, much as Aby Warburg had earlier brought Renaissance astrology into the history of art. The purpose of this chapter is to expose some fundamental presuppositions about the writing of the history of ancient science. In the case of Neugebauer and Pedersen, although they differed on the role of the Greeks in the invention of science, as well as on the question of rationality in cuneiform science, at the heart of both presentations lies a certain image of science's identity and purpose across history and culture. Neugebauer and Pedersen's historiographies of science, although for different reasons, line up on the same side of the so-called demarcation problem. In terms of comparative historiography, Neugebauer and Pingree exemplify different, even opposing, historiographical approaches. Chapter 3 focuses on the figures of Neugebauer, Pedersen, and Pingree because their work exemplifies distinct ways to approach the historiography of science in direct relation to particular construals of science, and arguably, the relation of science to a concept of "world." Neugebauer and Pedersen each founded their histories on the assumption of one science and one world. Pingree opened his methodology to a fundamental cultural plurality about science and a critique of the unity of science as represented by that of the modern West (Cat 2021; Tahko 2021). The recent generation of scholars in the sources from the regions (Middle East, Persia, India, Greece, and Rome) that produced science in antiquity, in spite of differences one or another may have with Pingree in the interpretation or dating of specific texts, is the result of the cultural pluralism of science articulated by him.

Part II in four chapters discusses new directions for analyses of cuneiform sciences afforded by reflecting on worldmaking with its attention to ontologies and the problem of pluriworldism. Chapter 4 deals with the various threads from the history of philosophy, the philosophy of science, and anthropology that all address in one way or another the debate between worldviews (conceptual schemes) on the one hand and the plurality of worlds on the other. The thorny issue of conceptual and ontological relativism pertains particularly to the history of distant sciences if what we seek to understand is a broader cultural and conceptual framework for those sciences rather than a comparison of their knowledge to ours. Such comparisons have their use, but not if we want to dispense with modern measures of ancient (or otherwise non-Western) knowledge.

Chapter 5 is framed by issues of cultural, epistemological, and ontological likeness, difference, and pluralism at the heart of anthropology. It brings these ideas to bear on the cuneiform intellectuals' particular mode of engagement with the world of their understanding through classification and model-making. My use of classification and model-making as heuristics for probing cuneiform worldmaking aims to show that the historiography of science in the ancient Middle East is an essential component of a broader discussion of difference, diversity, and counterethnocentrism that began in the 1960s and 1970s and continues to this day. The goal of this chapter is to show by example the relevance for Assyriology of the anthropology of science, the idea of worldmaking, and the ontological turn.

Chapter 6 turns to elucidate some of the ways the cuneiform scribes described their world and, from those descriptions, examines the concept of "world" in the cuneiform corpus. The inquiry into these ways of description is meant as an alternative to the reconstruction of a cosmology in the sense of a systematic account of the physical universe, such as is available from the well-known Greek treatises of Plato (*Timaeus*, see Bury 1929) and Aristotle (*De caelo*, see Guthrie 1971). Because the cuneiform evidence does not support such a reconstruction, it is argued that worldmaking serves better as a heuristic than does the idea of cosmology. Lastly, Chapter 7 looks at the question of the spherical world picture and the lack of evidence for such an image of the heavens or of the world as a whole in cuneiform sources.

In the history of the interaction of astronomy and cosmology, the premodern Western tradition is almost unthinkingly characterized as, in

Alexandre Koyré's words, "a finite and well-ordered whole, in which the spatial structure embodied a hierarchy of perfection and value ... unified only by the identity of its ultimate and basic components and laws" (1957, viii). Of course, this description does not fit every Greek cosmology in the run-up to Aristotle's *Physics* (Wicksteed and Cornford 1957), *Metaphysics* Λ (Frede and Charles 2001), and *De caelo* (Guthrie 1971). But ever since the decipherment of Babylonian astronomy, it has been plain that no such finite world with its basis in a physics of matter and motion can be assumed for the cuneiform world.

The proximity in geography of Babylon to Greece, and roughly in chronology from sixth-century BCE Babylon to the same period in which a Greek construction of the finite space of the cosmos is first found, not to mention the fact that the two cultures were aware of one another, makes a comparison between them seem obligatory, or inexorable. This comparison does more to show that a Babylonian cosmology in any sort of mode like that of the Greek did not exist.

The fundamental difference between Babylonian descriptions of the world and its parts, on the one hand, and the many discrepant Greek ideas within their philosophical tradition from roughly the sixth to the third centuries BCE, on the other, points to the real difficulty in generalizing about the modes in which ontological systems are expressed. In this context, I want to cite a remark by Carlo Severi (2013, in a review of Lloyd 2012) that "what we anthropologists tend to call 'cosmologies' are de facto regularities in the establishment of a number of shared assumptions, very rarely expressed in the form of an explicit argument, and always related to specific practices, systems of relationships, and genre of discourses, either ritual or mythological." With reference to Lloyd's discussion (2012, 60–61) of the many and irreconcilable ontologies in Greek cosmologies from the pre-Socratics to Aristotle, Severi characterized Indigenous cosmologies as "unsystematic," "philosophies *without an ontology*" in the Parmenidean sense (2013, 195; emphasis in the original). The search for a Parmenidean style of ontology in cuneiform texts also comes up empty.

RÉSUMÉ

An anthropology of the other and distant sciences necessitates our belief that it is possible, whatever our vantage point, to understand another point of view, to conceptualize the worlds implied in temporally remote and geographically distant sources, and to acknowledge them as ontological realities. It will in addition regard worldmaking as more than mere rhetoric and metaphor. From the anthropological point of view, to quote Benjamin Alberti: "By granting ethnographic subjects their own worlds ... we not only avoid explaining away non-Western alterity, we also unleash the power of ontological inquiry to push anthropological theory forward" (Alberti et al. 2011, 906). Adopting this point of view introduces a cultural-anthropological sensibility into the historiography of premodern science. Fortified with the philosophy of worldmaking, the "power of ontological inquiry" stands to push the history of premodern and non-Western sciences forward as well.

Assyriology's embrace of anthropology is not new. Already in 1960, A. Leo Oppenheim advocated anthropological methods for Assyriology in his essay "Assyriology – Why and How," published, not in the *Journal of Near Eastern Studies* or the *Journal of Cuneiform Studies* but in *Current Anthropology*. Later, this essay became the introduction to his *Ancient Mesopotamia: Portrait of a Dead Civilization* ([1965] 1977). There, he suggested that "if the new directions [toward interpretive strategies taken from the social sciences, such as anthropology] here surveyed mean that Assyriology will eventually move away from the humanities and nearer to cultural anthropology, I shall shed no tear" (1977, 30). I. M. Diakonoff's "More on Assyriology" in the second issue of *Current Anthropology* (1961, 390–91) expressed a similar leaning toward social science methodologies for the interpretation of cuneiform texts. This book aims to move nearer to an anthropology of science as an as yet untapped method for pushing the historiography of the cuneiform sciences of *ṭupšarrūtu* in a new direction.

PART I

HISTORIOGRAPHY OF SCIENCE

The chapters of Part 1 explore the historiography of cuneiform science in Assyriological scholarship and how this development fits into the field of the history of science more generally. The history of the modern interest in cuneiform science began with the decipherment and analysis of Babylonian astronomy and the "exact sciences" in the nineteenth to mid twentieth century and then broadened its scope in the late twentieth and twenty-first centuries to include omens, astrology, medicine, and magic. Accordingly, phenomena such as omens, portents, and signs came to be recognized in the scholarship of the past roughly fifty years not only as a part of cuneiform scientific inquiry but as the centerpiece of cuneiform scholarly knowledge and science. Early in our historiographical tradition, Babylonian mathematical astronomy was considered a separate and intellectually more sophisticated endeavor as compared to the divinatory and magical sciences, but we now realize astronomy's integral connection to the same tradition.

The reassessment within Assyriology of the full range of cuneiform sciences has also afforded an understanding of that tradition independent of the measure of modern ideas and scientific norms, with consequences for our approach to the epistemological and ontological dimensions of the cuneiform evidence. Trends in the history, sociology, and philosophy of science since the rise of postpositivism in the 1960s have been equally influential for that project. Contextualism, both cultural and societal, epistemic pluralism and debates over scientific realism and ontological relativism have contributed and continue to contribute to our ability to situate cuneiform sciences in the broader premodern epistemological landscape and a broader regional and chronological scope for a history of early science as a whole.

1

These Bones Live!

> Whether I have succeeded ... in making some of these "dry bones" live again, is up to the reader to decide.
>
> – A. Leo Oppenheim, 1967

The intersection of the field of Assyriology with the history of science is where three distinct but interconnected developments met and continue to interact. The first was the entry of Europe into the Middle East during the nineteenth century and the consequent birth of Assyriology, including the decipherment of Babylonian mathematical astronomical tables; second was the expansion of research on related materials for reconstructing the cuneiform scientific culture; and third was the consequent reassessment of the historiography of science with regard to Middle Eastern antiquity. The first has to do with nineteenth- and twentieth-century politics, which shaped and will continue to shape our knowledge of the ancient Middle East for as long as there is European involvement in the region. The second has to do with the recovery of an entire corpus of historical materials for the history of science in the region of the Middle East. The chronological scope of this source material spans millennia, with deep roots in the second millennium BCE and only disappearing in the first century of our era (CE), at least as far as extant cuneiform scientific texts are concerned. The sources for the intellectual and scholarly activities of the scribes are numerous, and their study is ongoing. The third intersecting point is situated at the place where the long history of the historiography of science meets cuneiform sources. This began to take shape in the eighteenth century when Europeans became aware of Babylonian astronomy as an ancient Oriental science (Swerdlow 1993). It continued in the nineteenth century with the discovery of the cuneiform astronomical tablets and the decipherment of Babylonian astronomy and was followed by the subsequent and continuing Assyriological study through the twentieth and into the twenty-first centuries of these and other related texts constituting the sciences of the scribes.

As argued in the Introduction, the term science justifiably applies to many of the activities within the epistemological manifold of the learned masters and *eruditi* among the cuneiform scribes. The scientific culture of the scribes, as will be detailed in this chapter, had a particular scope and nature to be differentiated from later, especially Western, scientific cultures. One of the key factors in reconstructing the cuneiform sciences from our present vantage point, however, is that even under the best of circumstances, the excavated remains of Middle Eastern antiquity represent a mere fraction of what once was a historical reality. Things have not substantially changed since A. Leo Oppenheim pessimistically stated:

> The cuneiform texts have given us a strangely distorted picture of more than two thousand years of Mesopotamian civilization. This picture is composed of abundant but very spotty detailed information, and of rough and incomplete outlines of the major political and cultural developments. All this theoretical framework, moreover, is torn to shreds again and again by immense gaps in time and space . . . ever-recurring blackouts of knowledge. (Oppenheim 1977, 11)

The frequency and magnitude of those "blackouts of knowledge" caused Oppenheim to ask, "*Can these bones live?*" (Oppenheim 1967, 56). I propose here that they can and they do.

EUROPE IN THE MIDDLE EAST AND THE DECIPHERMENT OF CUNEIFORM ASTRONOMY

From the beginnings of Europe's rediscovery of Middle Eastern antiquity in the period of British and French colonial interest in the region to the dark days of rampant looting following the American invasion of Iraq in 2003 and the iconoclasm of antiquities by ISIS in 2015 (Tugendhaft 2019, 2020), our formation of knowledge about the history of the region is inseparable from present concerns. The political, mostly predatory, relationship of Western nation-states to the Middle East has always surrounded archaeological research, compounding and complicating its sheer physical difficulty.

European politics has been and still is present in the reclamation and formation of knowledge about the diverse cultures of the Middle East. There has always been more than an abstract interest in the history of the region because we in the West are the heirs of ancient Middle Eastern cultures, not only of the Hebrew Bible (or what the Jesuits called the Old Testament) but also of the Babylonian, Assyrian, Egyptian, and Sumerian cultures, the oldest continuous literate cultures in human history. The impetus for rediscovery of Assyria and Babylonia in the mid nineteenth century was tied in part to the question of the identity of Judeo-Christian origins, in part to the activity of Jesuit scholars (Winitzer 2025). Recovering the history of the lands of the ancient Middle East, especially the place ancient Greeks called Mesopotamia (Μεσοποταμία),[1] held the potential to ground the Bible in historical reality in a way it could not do by itself. Its rediscovery would represent knowledge of a new antiquity materially relevant to European cultural and historical origins. Because the British and French governments were heavily invested in the region during the mid nineteenth-century partitioning of the Ottoman Empire, local European consuls and vice-consuls, together with officers of the British East India Company, as well as various travelers and businessmen, were on site poised for the study, excavation, and decipherment of cuneiform antiquities and languages.

The outrage felt in the modern West at the ransacking of the National Museum of Iraq in Baghdad and the loss of countless artifacts during April 2003 and beyond testifies to the connection felt to these profound symbols of our common heritage. And yet the cuneiform languages and scripts are alien, as are their modes of expression in subjects otherwise familiar to us, such as law, religion, mathematics, or science. Ancient Babylonia and Assyria, remote yet ancestral, bear an alien character stemming directly from the cuneiform remains themselves, the numerous clay tablets whose script had to be deciphered and whose texts penetrated, a process that is incomplete and ongoing today.

A different picture of an ancestral ancient Middle East was just as readily coopted by Saddam Hussein as he set about restoring the site of Babylon in 1978. The restoration included the creation of Arabic

[1] This ancient Greek term did not denote the territory the term "Mesopotamia" conventionally refers to today, and I prefer not to employ it. The history of the Greek term seems to begin with Arrian in the second century CE, where in the *Anabasis* VII, vii, 3 and the *Indica* 42:3 it is defined as an Alexandrian administrative district carved from the former Achaemenid satrapies of Ebernari and Babylonia (Finkelstein 1962, 73). Because of the predominantly cultural-historical thrust of my discussion, the term "cuneiform world" replaces "Mesopotamia" as a general designation for both the cultural and geographical range from which we have available historical sources written in various cuneiform languages. The cuneiform world represents a somewhat larger geographical realm than that covered by today's conventional use of the term "Mesopotamia."

inscriptions commemorating Saddam in the same formulaic language used by ancient Babylonian kings and the installation of these fabrications in the reconstructed walls of the palace of Nebuchadnezzar (Finkel and Seymour 2009, 213–20). To be sure, Saddam Hussein did not install these inscriptions or put his own image next to that of King Nebuchadnezzar on a coin because he thought Babylonia was the great symbol of Western European antiquity. What modern Iraqi and modern Western attitudes toward ancient Iraq have in common is the idea of a lineal relation between Iraqi antiquity and the present and, therefore, of heightened meaning for us, whoever we are (modern Europeans, Americans, or Iraqis), and for whatever reason.

Today, the prospect of furthering our knowledge of the ancient Middle East by continued excavation is threatened by the grave consequences of long-standing violent conflict and tensions in the region. The use of ancient sites as defensive military bases by the US military during the Iraq War, the spread of looting (from the 2003 looting of the National Museum of Iraq in Baghdad, the Sumerian site of Tell es-Senkereh/ancient Larsa, to other widespread damage to sites), and the wanton destruction by ISIS of Assyrian antiquities (Tugendhaft 2020) have all had an impact on research.

Less than 200 years after the rediscovery of the ancient empires of Assyria and Babylonia, some of what was reclaimed has already been erased. The loss of so much cultural heritage with its potential to enrich our understanding of the Middle Eastern past is catastrophic, but this loss is secondary to that of so much human life. Nevertheless, in the face of regional tensions and any ongoing potential for the erasure of history, the importance of the legacy of the cuneiform world for human history and the modern humanities should ensure its continuation – not as a mere academic luxury but as a valued necessity.

The ambiguous, that is to say both non-Western and Western, position of the cuneiform world is nowhere better represented than in the history of its science, particularly of the Babylonian celestial sciences. The point of entry at the intersection of Assyriology and the history of science occurred early in the development of Assyriology, when clay tablets found at Babylon and Uruk in southern Iraq became available in the West, mostly through unscientific excavation and acquisition through the antiquities market. These beginnings trace back to the early years of the nineteenth century, when on the strength of the young Englishman's prodigious linguistic abilities in Greek, Latin, Persian, Turkish, Arabic, Hebrew, and Syriac, Claudius James Rich was appointed assistant to the British consul general in Alexandria. In a few short years, he was moved to Baghdad as resident of the East India Company, and, in the following six years, Rich explored and wrote largely geographical and prearchaeological accounts of the British colonial region of what was once ancient Sumer, Akkad, Assyria, and Babylonia. The year 1811 saw the publication of his popular *Memoir on the Ruins of Babylon*, followed in 1818 by the *Second Memoir on Babylon*, only two years before his death at the age of thirty-three.

Claudius Rich opened the door and ushered in the brilliant and rapid period of decipherment of cuneiform and the identification of the languages of Akkadian and Sumerian during the middle decades of the nineteenth century. By that time, Assyrian royal inscriptions from the walls of palaces were being excavated in Nimrud, Nineveh, and Khorsabad. The famous Akkadian literary epic of Gilgamesh as well as Sumerian texts of many genres, literary and nonliterary, from southern Babylonian sites such as Lagash, Ur, and Uruk, had also come to light and were of interest to some of the original decipherers of cuneiform, such as Henry Creswicke Rawlinson and Edward Hincks.

As compared with the Classical Greek world, the cuneiform world at the time of its discovery offered new and hitherto-unexplored historical territory. Even though well-educated colonial agents of foreign governments may have been versed in the Bible and *The Histories* of Herodotus and could have been able to read Greek, Latin, and possibly Hebrew, as a matter of firsthand documentation, Iraq and its surrounding areas offered new possibilities for assessing the biblical and Classical narratives. Because the lands of the ancient Middle East (Sumer, Akkad, Babylonia, Assyria, Elam, Persia, the kingdom of the Hittites, to limit the list to cuneiform cultures) were previously known to Europeans only through the lens of biblical and Classical writers, the mid nineteenth-century decipherment of the cuneiform script opened a door to these ancient traditions without the filter of the Bible or the Greek historians.

Then toward the end of the nineteenth century, something unforeseen and unexpected came to light among the cuneiform tablets from Babylonia, namely ephemerides of the moon and the five naked-eye Classical planets, Saturn, Jupiter, Mars, Venus, and Mercury, that were not derivative of other ancient forms of astronomy known (Swerdlow 1993, 309–11). The realization of what these tables of cuneiform numbers represented was the result of the collaboration between an Assyriologist, J. N. Strassmaier, and a mathematician and astronomer, Josef Epping. By the turn of the twentieth century, F. X. Kugler in *Die Babylonische*

Mondrechnung (1900) had penetrated the Babylonian lunar theory, exploding any presupposition, widespread at that time, about the inability of so-called Oriental cultures to produce science.

The study of cuneiform astronomical texts began in the 1880s, when Epping and Strassmaier first revealed that the numerical table texts written on cuneiform tablets were lunar and planetary ephemerides (see Figure 1.1) (Epping and Strassmaier 1889). This revelation had a certain gravitas, because the tables analyzed by these pioneer scholars of Babylonian astronomy could be recognized as the oldest mathematical astronomy, the oldest exact science. As Otto Neugebauer pointed out:

> Epping fully realized the significance of his discoveries. The two columns from a lunar ephemeris which he had deciphered, he said, "give us more information about Babylonian science than all the notices from classical antiquity combined" – a fact which cannot be emphasized too often. And he [Epping] foresaw clearly that the new material would become of great importance for ancient chronology, for Assyriology in general, and even for modern astronomy. (Neugebauer 1975, 349 and note 6)

During the very period of the recovery and decipherment of the cuneiform astronomical texts, another scholarly movement was under way that would directly relate to the eventual incorporation of the new field of Babylonian astronomy and astrology into a deeper understanding of the astral sciences of the entire ancient Mediterranean and Middle East. A contemporary of Epping, Strassmaier, and Kugler, the Belgian Classical philologist and historian Franz Cumont, together with Classical philologists Franz Boll and Wilhelm Kroll, was engaged in what would ultimately be the twelve-volume *Catalogus Codicum*

FIGURE 1.1 Babylonian lunar ephemeris, from Otto Neugebauer, *Astronomical Cuneiform Texts* (London: Lund Humphries, 1955), nr. 122 (BM 34580). Photo courtesy of the Trustees of the British Museum.

Astrologorum Graecorum (CCAG, see Cumont, Boll, and Kroll 1898–1953).

The collection of the Greek astrological texts would open new possibilities for the study of how astronomy and astrology were interdependent and how the astral sciences functioned within the *oikoumene*, including, of course, the cultural-geographical area of the ancient Middle East (and beyond). In 1911, for example, in the *Sitzungsberichte* of the Heidelberg Academy of Sciences, Boll, together with Semitist and Orientalist Carl Bezold (Bezold and Boll 1911), set out extensive parallels between the then newly available cuneiform celestial omen texts and certain Greek works from the CCAG as well as, for example, the sixth-century CE work of John the Lydian, or "Lydus," on divination, titled *De Ostentis*, "On Signs." This material was proof of an extensive transmission of Babylonian astronomical knowledge, a phenomenon that would occupy many historians of Babylonian astronomy throughout the twentieth century, such as Otto Neugebauer and David Pingree (Misiewicz 2019).

Roughly half a century after the foundation was laid by the Jesuits Epping, Strassmaier, and Kugler for the field of Babylonian astronomy (de Jong 2016), Neugebauer (1955) brought out a critical edition, with transliteration, translation, and commentary of the entire corpus of cuneiform lunar and planetary tables and procedure texts from Babylon and Uruk of the fifth to the first centuries BCE. This work, *Astronomical Cuneiform Texts* (ACT), is still a cornerstone for the field, now supplemented by Ossendrijver 2012 and 2024. In ACT, the two basic calculation methods discovered by Epping, Strassmaier, and Kugler, coined by Neugebauer as Systems A and B, were elucidated. This publication must be seen against another herculean effort to publish astronomical and astrological cuneiform texts in hand copy, for example, Sachs 1955 (with Pinches and Strassmaier = LBAT) and the monumental Sachs–Hunger–Steele 1988–2022 (ADART). In addition to the exposition of Babylonian astronomy in Neugebauer 1975 (HAMA), important analyses of Babylonian lunar theory (Brack-Bernsen 1969, 1980, 1990, 1997, 1999; Britton 1987, 1989, 2002, 2007b, 2009) and planetary theory (Swerdlow 1998) are found in the secondary literature.

The recovered astronomical cuneiform texts would ultimately change the face of the history of astronomy and, by extension, the history of science itself. Neugebauer's three-volume *A History of Ancient Mathematical Astronomy* (1975) placed Babylonian astronomy firmly in line with the tradition of Ptolemy's *Almagest* and all later Western astronomy up to Copernicus. Neugebauer credited to F. X. Kugler (Neugebauer

1975, 305–6) the discernment of Ptolemy's debt to the Babylonians underlying the Hipparchan lunar parameters used in the *Almagest* (IV.2, Toomer 1984, 174–79), specifically from the lunar System B (see Glossary of Astronomical Terminology Cited). The recovery of the bones of Babylonian astronomy made it possible to trace survivals of its parameters and methods not only in Greek but also in Indian and medieval European astronomy.

After Neugebauer, the direct link from Babylon to the West through the transmission of astronomical knowledge[2] to Greece and the Greco-Roman world would come to occupy a central place in assessing the relation of Babylonian knowledge to later science. The impact of the initial decipherment and later explication of cuneiform astronomical texts on the historiography of science of the late nineteenth and early twentieth centuries, therefore, had explosive potential because the most entrenched idea about the history of science of that entire era was the idea that science originated with the Greeks. This potential was a long time in coming, as various arguments were put forward to explain and justify the claim to the Greek invention (critiqued in Rochberg 2004, 14–43) even after Babylonian astronomy was a known quantity, at least to specialists.

One example, from 1954, the year before the appearance of Neugebauer's ACT, is found in Erwin Schrödinger's book *Nature and the Greeks*. In the chapter titled "Return to Antiquity," he quoted Theodor Gomperz, a somewhat older contemporary of Kugler, from his work *Griechische Denker*, first published in 1893 and in its third edition in 1911, still relevant for Schrödinger and his audience in the mid 1950s:

> Nearly our entire intellectual education originates from the Greeks. A thorough knowledge of these origins is the indispensable prerequisite for freeing ourselves from their overwhelming influence. . . . Not only has their [Plato's and Aristotle's] influence been passed on by those who took over from them in ancient and in modern times; our entire thinking, the logical categories in which it moves, the linguistic patterns it uses (being therefore dominated by them) – all this is in no small degree an artefact and is, in the main, the product of the great thinkers of antiquity. (Gomperz, within Schrödinger 1996, 19–20)

The salient point about the Greek invention of science was that it inaugurated a particular kind of thinking – "our entire thinking," as Gomperz said, implying all forms of rational thought. This quality of

[2] By "astronomical knowledge," I refer to all forms of knowledge of the heavens and heavenly phenomena in antiquity, including technical astronomy, astrology, and all related interests in the phenomena.

mind was long and widely regarded as the special possession of the West. The fact that Babylonian astronomical ideas and parameters enabled the development of Greek mathematical astronomy, a historical fact known by 1911 when Gomperz wrote and well known by the mid twentieth century, was still not seen as in any way part of the history of "thought."

Today, the rhetoric of a Greek monopoly on rational scientific thought in antiquity has an essentialist, crude, and artificial ring to it. Thanks to Neugebauer and his Brown University colleagues' attention to sources outside of the Greek corpus, a better way of understanding the complexities of the culture, or the cultures, of astronomical science in the Hellenistic world could be considered. The study of the non-Greek sources for the astronomical sciences – within which I include observational, predictive, and mathematical astronomy, genethlialogical astrology, and celestial divination – in Babylonian, Egyptian, Judean, and Indian texts showed that traditions coexisted and were transmitted, received, adopted, and reformulated. In other words, the "Greek way" of thinking about science was itself, in no small measure, formed by contact and exchange with cuneiform and other cultures that Greek intellectuals encountered through the political and cultural world established after Alexander's conquests.

Even though early twentieth-century historiographies of science were fraught with prejudice against "Orientals" and "primitives" (i.e., non-Greek ancient peoples), the original cuneiform astronomical texts made it clear that Greek astronomy did not spring as Athena full grown from the head of Zeus, but itself owed a sizable debt to Babylon. The claims about Chaldean astronomy found in Greco-Roman sources such as Geminus, Ptolemy, Pliny, Diodorus, and others could finally be assessed against cuneiform texts, and a basis for comparison was thus established. Once one took account of the units (sexagesimal numbers, the 360-degree circle, the cubit, and the finger), observations (e.g., lunar eclipse observations given in Ptolemy's *Alm.* IV.6, Toomer 1984, 191–92), parameters (one of the famous Babylonian parameters is the length of the mean synodic lunar month as 29;31,50,8,20 days in the lunar System B), and period relations (the 19-year lunisolar cycle also known as the Metonic cycle in which 19 sidereal years = 235 lunar synodic months; the Saros cycle to predict eclipses for which 38 eclipse possibilities = 223 synodic months) adopted from Babylonia by Greek, Greco-Egyptian, and Greco-Roman astronomers and astrologers, it became clear how extensive the Babylonian contribution to Hellenistic astronomical science in fact was.

Where Babylon had influenced Greece, a greater relevance or legitimacy could be attributed to the Babylonian tradition by virtue of its making the advances of Greek science possible. This was of a piece with other aspects of cuneiform culture, its urbanism, law codes, and well-developed military capacities, which were viewed as continuous with and contributing to the construct of "Western Civilization." Thus, the Fertile Crescent came to represent the "Cradle of Civilization," where civilization is synonymous with that of the West. Speaking from a broad historiographical standpoint rather than specifically about science, Marc van de Mieroop observed that "the predilection to see the Ancient Near East primarily as a precursor of the Judeo-Christian and Graeco-Roman legacy tacitly presents the European cultural development as the superior one in the world and measures the relevance of other traditions only in relationship to it" (van de Mieroop 1997, 288). Similarly, insofar as cuneiform science anticipated Greek developments, it took its place in the history of science.

The increasing evidence of the importance of Babylonian astronomical sciences to the Greeks, Romans, Judeans, and Indians spearheaded a major effort to trace the transmission of Babylonian knowledge to these other cultures. The work to trace Babylonian number-notation style, parameters, methods, and schemata to other cultures began in 1911 (Bezold-Boll 1911), was expanded and deepened by David Pingree (Pingree 1997), and continues to this day (Brown et al. 2018; Misiewicz 2018). Not only is van de Mieroop's observation, therefore, a critique of historiographical teleology because it can result in assessing earlier traditions as less developed and less sophisticated and therefore lesser in all respects than what comes later, it is also an invitation to take the cuneiform sources on their own terms. Although science was not part of van de Mieroop's remit, the question about teleological historiography was clearly fraught for historians of science.

ṬUPŠARRŪTU AND ITS SCIENTIFIC CULTURE

The second of the elements at the intersection of Assyriology with the history of science is that of our developing study of the cuneiform scientific culture itself. Taken as a totality, the sciences of the cuneiform world of ca. 2000 BCE to ca. 100 CE, including divination, astronomy, astrology, magic, and medicine, have enormous significance for the historiography of science. Their significance is due to the unique combination of what we can identify as points of contact with conventional ways of recognizing science as well as presenting a radical otherness in other respects.

The sciences in question comprise the knowledge corpora and associated practices of *ṭupšarrūtu*, the term for the component scribal scholarly disciplines that organized knowledge of the phenomenal world and the practices that depended upon that organization. Textual evidence for *ṭupšarrūtu* is extant from the Neo-Assyrian period (seventh century BCE) from such sites as Nineveh (Kuyunjik), Assur (Qalʿat Sharqāṭ), Kalhu (Nimrud), and Huzirina (Sultantepe) to the Neo-Babylonian and Seleucid periods (fifth to second centuries BCE) from sites such as Babylon and Uruk (Robson 2019, 52–53). Arguably, even earlier texts from Old Babylonian periods (including Hittite texts) belong to this tradition, but the term *ṭupšarrūtu* itself is a designation found only in first-millennium Akkadian sources.

Morphologically an abstract Akkadian noun from the professional designation "scribe" (DUB.SAR = *ṭupšarru*), *ṭupšarrūtu* is defined (CAD, s.v. meaning 2) as "scribal learning, scholarship." The forms of scribal scholarship encompassed by the term *ṭupšarrūtu* produced a distinct textual and intellectual culture. Moreover, in *ṭupšarrūtu* we see the marks not only of a textual and intellectual culture but also of a scientific culture.[3]

In addition to the rubric *ṭupšarrūtu* "scribal knowledge" for this literature and its traditions, there were other Akkadian terms such as *nēmequ* "skill" or "technical knowledge," as well as other less well-attested terms for knowledge, scholarship, and expertise in skills, crafts, and artisanry (*ummânūtu*, *mudûtu*, *lē'ûtu*). The term *nēmequ*, from a root meaning "to be able," connotes knowledge associated with the production of various material technologies (Escobar 2017, 17, 19–20, and throughout).

Expertise and knowledge (*nēmequ* and *ṭupšarrūtu*) were closely connected, but these terms could refer in some cases to different textual domains. In other cases, the terms seem to be mutually inclusive. At the height of the Assyrian empire, for example, a prism inscription describes how King Assurbanipal was educated in the "house of succession" (*bīt ridûti*):

> I, Ashurbanipal, learned inside it [the bīt ridûti] the expertise (nēmequ) of the god Nabû, the entire corpus of scribal scholarship (kullat ṭupšarrūti). I probed every kind of knowledge of every kind of scribal master in existence.[4]

It is clear from this passage that *nēmequ* connoted skill and mastery of technique, in addition to knowledge of "the entire corpus of scribal

[3] The relationship between *ṭupšarrūtu* as cuneiform knowledge and our term "science" is discussed in Rochberg 2016a, 9–10, 34–35, 61–102; Robson 2019.

[4] Prism F i 24–26, http://oracc.museum.upenn.edu/rinap/rinap5/pager.

scholarship." We learn from this same passage that the knowledge in which the crown prince Assurbanipal was educated included equitation ("how to shoot a bow, ride a horse [and] chariot, [and] take hold of [their] reins," lines 27–28) and the art of warfare. The skill of writing itself was called the "*nēmequ* of Nabû (the patron deity of writing)." Thus, *nēmequ* has particular connotations that the word "wisdom" in later Western contexts does not have. The clarification and definition of *nēmequ* and how it relates to our word "wisdom" is an important complement to the recent redefinition of science that takes into account categories and criteria at stake in original sources.[5]

In the colophons to scholarly texts stored in Assurbanipal's palace during the seventh century BCE, the tablets comprising the various fields of *ṭupšarrūtu* were described as *nisiq ṭupšarrūti* "the highest level of scribal scholarship," *nēmeq Nabû* "the wisdom/skill of Nabû, patron deity of writing," and *tikip sattakki* "the cuneiform signs." As an account for why the texts belonging to *ṭupšarrūtu* and *nēmequ* held the highest rank, the masters (*ummânū*) of the fields of *ṭupšarrūtu* took their lineage from seven mythological antediluvian sages of the underground sweet waters where the god of wisdom and patron deity of the master scholars, Ea, had his abode. In this way, the uppermost echelon of scribes who wrote scholarly texts, including astronomical ephemerides, identified themselves with a tradition of knowledge they claimed was handed down to them from the remotest antiquity before the Flood. This tradition has been characterized as a "mythology of scribal succession" (Lenzi 2008, 109, 115, 119, and throughout) and is a significant element of how scholarship and knowledge were defined in cuneiform culture.

Learning fell under the patronage of the gods, expressed as *nēmeq Nabû* "wisdom/skill of Nabû" and *nēmeq Ea* "wisdom/skill of Ea," which is said of a scholarly tablet, and the scribe who wrote it was called "one who understood the entirety (*kullatu*) of *ṭupšarrūtu*" (Hunger 1968,

[5] Peter Harrison (2006, 51) traces "the gradual dissociation of wisdom from both theology and the natural sciences" over the course of the centuries from the early Christian fathers through the early modern rejection of Aquinas's views on knowledge, theology, science, and wisdom. According to him, patristic views opposed heavenly wisdom (about God) and earthly wisdom (about Classical science), a split that later came together when Aquinas counted the earthly wisdom of science as a path to the heavenly wisdom of theology. The medieval state of affairs is relevant to the present context of cuneiform science if only as a reminder that wisdom and science have been at times opposed, at times united, but after the cuneiform tradition, this relationship has always been gauged in terms of a relationship or opposition between natural and divine forms of wisdom. There was no separation along those lines for Babylonian scribes.

nos. 330:5, 331:6; both Assurbanipal palace colophons). Divine patronage of learning is seen in every corner of the texts comprised by *ṭupšarrūtu*. Judging by these expressions, throughout the Neo-Assyrian period divine patronage was mostly identified with Ea, the patron of wisdom and magic, but also the god Nabû, patron of writing, with his goddess Tašmētu (Robson 2019, 53–85). Rituals for the diviner (Starr 1983) who inspected the entrails (*bārû*) appealed directly to the divine patrons of divination, Šamaš and Adad, who were said to communicate their decisions by writing on the liver, sometimes called "the tablet of the gods" (Lambert 1998, 148–49: 8, 14, 16). Writing divinatory texts in cuneiform on tablets was by this metaphor elevated to divine status.

In the Seleucid period, the idea of divine wisdom and knowledge is also attested in Late Babylonian astronomical ephemerides. The contents of these tablets are sometimes described in colophons (Hunger 1968, no. 98; also in Neugebauer [1955] 1983, 18 as Colophon U), much as in the Neo-Assyrian colophons, as *nēmeq anūti* ("the wisdom/expertise of Anu-ship"). As *anūtu* is the abstract form of the divine name Anu, the divine head of the pantheon and god of the heavens, *nēmeq anūti* is the highest order of wisdom and knowledge/skill. *Nēmeq anūti* was also held to be a secret of the great gods, and the possession of the *ummânu*, the absolute scribal master of *ṭupšarrūtu*. On the upper edge of ephemerides from Late Babylonian Uruk, the sky god and his goddess, Anu and Antu, were regularly invoked (Bēl and Bēltīja in the texts from Babylon), also in a horoscope text (Rochberg 1998, text 14), with the formula *ina amat Anu/ Bēl u Antu/ Bēltīja lišlim*: "By the command of Anu and Antu (or Bēl/ Bēltīja), may it go well/remain intact." Whereas some of the astronomical literature, such as the astronomical observation texts now known as Diaries (in antiquity as the "regular watch," see ADART), does not appear to belong to this category of divine knowledge, both the exclusivity of access to divine knowledge and no doubt as well the highly technical nature of the bulk of astronomical texts made those texts fit to be seen only by the "knowledgeable" scribe (called *mudû*, Rochberg 2016a, 62).

For insights into the epistemic interests and methods of *ṭupšarrūtu*, the lists of text incipits in catalogues provide a bird's-eye view of the scholarly genres for which such catalogues remain (Frahm 2011a). Our grasp of the contents and structure of many individual compositions and compendia of cuneiform scribal scholarship owes much to the preservation of text catalogues that list the incipits of such works. Old Babylonian text catalogues may have originated as inventories of the holdings of particular collections (Delnero 2010) or as an ordering system for use in a scribal school curriculum (Tinney 2011). Features of the organization of those early literary catalogues carry over into the first-millennium scholarly inventories of *ṭupšarrūtu*, in particular of omen texts, seen in the sequences of thematically grouped texts, the use of subscripts to identify the type of text catalogued, and the variable inclusion of different text types within a single catalogue. First-millennium (Neo-Assyrian and Neo-Babylonian) catalogues are aimed either at a particular genre of texts, such as incantations, or at tablets within a discrete compendium, such as the celestial omen series *Enūma Anu Enlil*.

In the main, *ṭupšarrūtu* consisted of a wide variety of multitablet omen compendia. The omens compiled in these formalized text series (e.g., the series *Enūma Anu Enlil* comprised seventy tablets) were based on the observation not only of the details of human experience but also of terrestrial and celestial phenomena. Intrusions of one into the other may be found for all seven of the major compilations (as outlined in Rochberg 2004, 54):

- *Enūma Anu Enlil* ("When Anu and Enlil," the celestial omen series)
- *Šumma ālu* ("If a city," the terrestrial omen series)
- *Sakikkû* (omens devoted to symptoms of an illness, both prognostic and diagnostic)
- *Alamdimmû* ("If the form," the series for physiognomy and morphoscopy, with its poorly attested subseries *Nigdimdimmû* "If the appearance" and *Kataduggû* "If the utterance")[6]
- *Šumma izbu* ("If an anomalous birth," the series for omens from malformed fetuses and other irregularities of births)
- *Ziqīqu* (the series for dream omens)
- *Iqqur īpuš* ("He tore down, he rebuilt," the series for the propitiousness of dates for undertaking various activities or for someone born on certain dates)

These series compiled omens from so-called unprovoked signs, things that happen independently of the diviner's actions to "provoke" them. The omens resulting from the diviner's provocations were the consequence of actions that appealed to the gods, Šamaš and Adad, for an answer, providing them with a medium of communication, such as the

[6] An important discussion of the relations and connections among the series *Sakikkû*, *Alamdimmû*, *Nigdimdimmû*, *Kataduggû*, *Šumma sinništu* ("If a woman"), *Šumma liptu* ("If a spot [on the body]"), and even *Šumma ālu* is Schmidtchen 2018 and 2019, and for cuneiform physiognomic texts in general, Johnson and Stavru 2019.

sacrificed sheep, dropping oil into water, releasing smoke from a censer, or sprinkling flour. Of the provoked omens, extispicy (inspection of the entrails) had an extensive series for omens from the inspection of various entrails, such as the liver, gall bladder, intestines, and lung. The provoked omens came under the heading *barûtu*, meaning inspection by extispicy. Accordingly, the *bārû* ("diviner," literally "the one who makes an inspection") was the diviner specializing in provoked omens from the exta, oil, and smoke.

It is fortunate that the celestial omen series *Enūma Anu Enlil* is the subject of two extant catalogues, one from seventh-century Assur and the other from second-century Uruk (dated 26 Šabaṭu year 117 SE = 24/25 February 194 BCE). The Uruk catalogue gives the incipits for *Enūma Anu Enlil* Tablets 1–26. The Assur catalogue (excerpts given on pp. 46–7) contains incipits for *Enūma Anu Enlil* Tablets 39–59 and twenty-nine single-column *aḫû* tablets (IM.GÍD.DA.MEŠ BAR.MEŠ), meaning tablets with noncanonical content. Following the list of incipits for the celestial divination series are incipits for the terrestrial divination series *Šumma ālu* Tablets 1–22 and 30–62. Whatever else might have been a part of this list is unknown due to the broken lines at the end of the catalogue (rev. col. iv). The 400-year gap between the catalogues means that they display a number of superficial differences, but overall their function was the same.

The combination of the celestial with the terrestrial may be an important factor to take into account in analyzing the Assur catalogue, as it could reflect upon its purpose and how the scribes classified the materials of *ṭupšarrūtu*. The combination of other disparate scholarly series together in a single catalogue (such as the combination of diagnostic omens from the series SA.GIG, and physiognomic omens from the series *alamdimmû*, Finkel 1988) could shed some light on the practice in general and whether the Assur catalogue of *Enūma Anu Enlil* and *Šumma ālu* (and perhaps a third or more series) is unusual or not.[7]

In order to better convey the essence of the material described in this chapter, some lines from the Assur catalogue are given here (for details see the full edition, Rochberg 2018b). For comparison, a number of examples of omens from the tablets whose titles are entered in the catalogue are given as well. The importance of this material here is both in the content of

[7] The pairing of the celestial with the terrestrial is known from the Diviner's Manual (Oppenheim 1974), but the incipits of celestial and terrestrial omens given in the Diviner's Manual do not come from *Enūma Anu Enlil* or *Šumma ālu* and so remain an exceptional case.

the series catalogued and in the manner of description of the various phenomena.

Assur Catalogue obverse col. i

1′ [If Adad thun]ders [in the east.] Tablet 39.
2′ [If Adad t]hunders [in the middle of the Pleiades.] Tablet 40.
. . .
4′ [If an earthquake shak]es [in Nisannu.] Tablet 42.
5′ [If in Nisannu the south wind blo]ws. Tablet 43.
6′ [If the Wagon Star (is visible) for an ecl]ipse. Tablet 44.
7′ [If the Field is vis]ible [in Nisannu.] Tablet 45.
8′ [If the Field (γ Pegasi) and the Stars/Bri]stle (Pleiades) are visible. Tablet 46.
9′ [If the Stars/Bristle re]ach [the Yoke (Boötes).] Tablet 47.
. . .

obverse col. ii

1′ – 2′ [If (Jupiter)] approaches the [place] where the sun becomes visible and stands.
3′–4′ If on the 12th day of Nisannu the moon is early and becomes dark in the evening; his features are like sulphur fire.
5′ Total of 29 Tablets, alternative one-column tablets.

6′ Tablet of incipits of (the series) If a city is situated on high ground. Series not (completed?). . . .

7′ If a city is situated on high ground. [Tablet 1]
8′ If in the observation hole of the city gate.
9′ If a brick mold (or) blade of a spade.
10′ If during the destruction of his house.
11′ If the foundation of his house was laid on the 16th day.
18′ If [white] lichen [appears in] the house [of a man.]
. . .
19′ If fungus [appears] in the city square.
20′ If holes open up in houses and cities.
21′ If water is spilled at the door of a man's house.
22′ If a man thinks every day about making his grave.
. . .
25′ If a goat-like demon is seen in a man's house.
26′ If a flash of light is seen in a man's house.

27′ If a dead man like a living one is seen in a man's house.
28′–29′ [If] on the first day [in the month Nisannu] a man, before he even puts his foot (out of bed) [onto the ground] sees a snake.

This excerpt from the Assur catalogue of incipits from two of the principal omen series, celestial and terrestrial, offers a glimpse of the kinds of phenomena identified by the scribes as ominous. Although this particular selection seems to deal with observable things (with the exception of the incipit concerning the sighting of a goat-like demon), the omen texts themselves contain many statements in which a phenomenon that is said hypothetically to be "seen" is by our lights not possible to see. Observable and unobservable, material and immaterial, physical and nonphysical things alike are identified as celestial or terrestrial phenomena (Rochberg 2009a, 16–20, 2010b). We must resolve this divergence from the physical phenomena we define as objects of scientific inquiry by investigating the aims and the conceptual frameworks of this and other corpora of cuneiform knowledge.

Apart from the vast collection and systematization of omens and their different series, *ṭupšarrūtu* also encompassed the sciences of astronomy and medicine. What we call astronomy consisted of a number of well-defined genres of such texts devoted to astronomical observation, schematization, and nonmathematical and mathematical prediction (Hunger and Pingree 1999 provides a survey), including horoscopes (Rochberg 1998). What we call medicine consisted of a number of interrelated and interdependent forms of the science of healing, namely, *āšipūtu* (knowledge and practice of conjuration against evil, and incantation and prayer literature) and *asûtu* (medical practice and knowledge of medicines, as in Geller 2010; Schwemer 2019, 39–41).

The *āšipu* was expert in techniques of appealing to the gods to heal the sick, such as incantations and rituals for ridding the patient of whatever consequences he would suffer from bad omens (*namburbû*), especially those responsible for illness. This expert did not simply come in after diagnosis to heal through ritual and incantation but was a master of the medical diagnostic omen series *Sakikkû* and the physiognomic series *Alamdimmû*. Together, these omen compendia combined knowledge of symptoms, diagnostics, prognoses (of recovery or death) in the case of certain symptoms, and all the anatomical regularities and irregularities of the human body.

The *āšipu*'s colleague, the *asû*, specialized in the practice of administering medicine in the form of drugs, the many preparations made from a wealth of *materia medica*, as well as the use of bandages. Just as those of the omen series *Enūma Anu Enlil* and *Šumma ālu*, the texts of *asûtu* were catalogued in what is now called the Assur Medical Catalogue (Steinert 2018, 11, 13–14, 172–84, and [with S. V. Panayotov, M. J. Geller, E. Schmidtchen, J. C. Johnson] 203–91 [for the edition of the text]; Panayotov in Steinert 2018, 89–120). As Geller and Steinert showed (Geller 2010, 9; Steinert 2018, 187–92), there was considerable overlap between the two kinds of medical practice, while they were nonetheless internally classified under two rubrics (*āšipûtu* and *asûtu*). Thus, the separation of the two into medicine (*asûtu*) and magic (*āšipūtu*) as though these distinctions parallel our own separation of medicine from "alternative medicine" makes for a false dichotomy and a misclassification of the evidence.

A list of texts in the corpus of *āšipūtu* "the art of the incantation expert" covers the text series and subjects to be mastered by the *āšipu*, a priest whose duties included the conjuration of demons for the purpose of healing the sick and also diagnoses of illness. The text in question (KAR 44, Geller 2018, 292–312) opens with "The incipits (i.e., titles) of exorcism compositions, established for study and reading (lit. 'viewing'), named in their entirety" (Geller 2018, 297) and is then followed by a number of omen texts also belonging to the *āšipu*'s repertoire, namely, *Sakikkû*, *Alamdimmû*, *Nigdimdimmû*, and *Kataduggû*. Further incantations, purification rituals, prayers, and spells are also listed, as well as "predictions from stars, birds, oxen, and flocks, oracles (based) on stones (or) flour, on incense, (and) on a god, in their totality, 'explanatory stone lists,' 'explanatory plant lists,' the 'tablet of stones,' the 'tablet of drugs,' 'strings' and 'pendants'" (Geller 2018, 299–300).

Celestial and terrestrial omens together with knowledge of medicinal plants, amuletic stones, and incantations and apotropaic rituals, together with omens from inspection of the liver and exta of the sacrificed lamb, as well as mathematics and astronomy all belonged to *ṭupšarrūtu*. Nearly this entire range of subjects was claimed as the extensive learning of King Assurbanipal, as follows:

Marduk, the sage of the gods, gave me wide understanding and broad perceptions as a gift. Nabû, the scribe of the universe, bestowed on me the acquisition of all his wisdom as a present. Ninurta and Nergal gave me physical fitness, manhood and unparalleled strength. I learnt the lore of the wise sage Adapa, the hidden secret, *the whole of the scribal craft* (*kullat ṭupšarrūtu*). I can discern celestial and terrestrial portents and deliberate in the assembly of the experts. I am able to discuss the series "If the liver is a mirror image of the sky" with capable scholars. I can solve convoluted reciprocals and calculations that do not come out evenly. I have read cunningly written text in Sumerian, dark Akkadian, the interpretation of which is difficult. (K 2694 + 3050, from Livingstone 2007, 100, lines 10–18; my emphasis)

The totality of scholarly fields and the many phenomena and areas of interest pertaining to the sciences of *ṭupšarrūtu* make very clear that a historiography of science reducing the aims and characteristics of science to those that stem from a modern sensibility about science, that is, one aimed at discovering and representing the physical workings of nature, will not suffice. Some of the premodern sciences, such as Ptolemaic and Copernican astronomy, medieval natural magic, alchemy (as well as metaphysics), and Renaissance astrology have already played a role in a reappraisal of the scientific revolution (Lindberg and Westman 1990). The more remote and distant sciences of *ṭupšarrūtu* present another and somewhat different opportunity for a reassessment of the meaning of science in historical as well as cultural contexts.

Whether it is best to speak of *ṭupšarrūtu* as the reflection of an intellectual culture (Stevens 2019), a textual culture, or a scientific culture seems to me a function of the questions one poses. In my view, it encompassed all three. The corpora of *ṭupšarrūtu* constituted an intellectual culture by virtue of being the product of an educated elite. The high status of this elite community of scribes, regardless of whether they were court appointees in the palaces of Assyrian kings or specialized priests and experts employed by the Babylonian temple during the Neo-Babylonian and Seleucid periods, remained consistent over time. As a textual culture, *ṭupšarrūtu* shared some features with medieval textual cultures, such as are discussed in Wallis and Wisnovsky (2016). The amalgam of traditions from the entire Greco-Roman realm, including the Arabic tradition of learning and science, ultimately inherited some of the traditions of cuneiform cultures in the fields of astronomy, medicine, and divination, in short the scientific fields from the second century BCE. That the Hellenistic/Greco-Roman intellectual culture included those scientific cultures from the East has long been known, even in antiquity. Furthermore, the unity of *ṭupšarrūtu* is also suggested by the relationship its series had to secret knowledge (discussed in Rochberg 2004, 214–17 and Lenzi 2008, 143).[8] Finally, if we define science in terms of the practices and

[8] Alan Lenzi said, "Late second and early first millennium sources on secrecy and scribalism use a word familiar to this study to describe the scribal craft: *niṣirtu*. The excursus to chapter one of this study [Lenzi 2008] noted the semantic proximity of *niṣirtu* and *pirištu* based on a synonym list (Aa = nâqu II/4 52–53). Interestingly, the very next word in this list is *ṭupšarrūtu*, 'the scribal craft.' This list, it seems, sets the three terms into a close semantic

methods of inquiry into and the resulting systems of knowledge of what a particular community imagines and perceives as its phenomenal world, *ṭupšarrūtu* is the instantiation of science in the cuneiform world.

Most importantly, how the cuneiform scribes defined and understood the phenomena, the objects of their inquiry, is the crux of how they understood their world. How the objects of the scribes' inquiry were understood, then, is a central question for both historical epistemologies *and* ontologies. The kind of knowledge encompassed by science and the relation it has to its world underscores the integrated nature of epistemologies and the ontologies supporting them, regardless of cultural or historical context. This aspect, of ontology and worldmaking, is reserved for Part II of this book.

Finally for this chapter, the following section takes up the third intersection point of Assyriology and the history of science in very broad outlines. Chapter 2 develops this historiographical context further.

HISTORIOGRAPHY AND CUNEIFORM SCIENCE

The third point of intersection of the field of Assyriology with the history of science was made possible by a new direction in the historiography of science associated with Kuhn in the 1960s (Oliveira 2012; Daston 2016) but presaged earlier, especially by Koyré in the 1950s (Pisano, Agassi, and Drozdova 2018). This development, in bringing matters of social and other contextual factors to bear on the study of science in history, would ultimately open new possibilities for a different approach to cuneiform science within a broader methodological scope. This postpositivist history of science had a decisive impact on my own approach to cuneiform knowledge in relation to science.

The sociology of scientific knowledge that emerged in the 1960s and 1970s, by grounding science in its social and historical contexts, provided an argument for the historicity of epistemic categories and a consequent pluralism applied to standards of scientific knowledge. The alternative to the so-called internal history of science as the story of progress and cumulative knowledge detached from history and social factors was a so-called external historiography, in which the social dimension of science and its practice took center stage. Eventually, however, it would be recognized that, strictly speaking, there is no division between internal

relationship. If there is evidence for attaching secrecy to the scribal craft in general, this text, originating in the 12th century, is the first glimpse of it."

and external histories of science once the nature of science as a cultural phenomenon is fully acknowledged.

Although contextualism for science in history was not entirely new, Kuhn's *Structure of Scientific Revolutions* ([1962] 1996) was the book that galvanized the history and philosophy of science around the historical, cultural, and social dimensions of scientific knowledge, standards, and norms. In the opening chapter, titled "Introduction: A Role for History," Kuhn stated his aim as "a sketch of the quite different concept of science that can emerge from the historical record of the research activity itself. Even from history, however, that new concept will not be forthcoming if historical data continue to be sought and scrutinized mainly to answer questions posed by the unhistorical stereotype drawn from science texts" (Kuhn [1962] 1996, 1). Kuhn's approach had a complex development, outlined in detail by Oliveira (2012). Oliveira's tracing of the new historiography of science looks at change from inside as well as out. Regarding the endogenous shift in the historiography of science, he (Oliveira 2012, 116) quoted the following passage from Kuhn's *Structure*:

> Gradually, and often without entirely realizing they are doing so, historians of science have begun to ask new sorts of questions and to trace different, and often less than cumulative, developmental lines for the sciences. Rather than seeking the permanent contributions of an older science to our present vantage, they attempt to display the historical integrity of that science in its own time. (Kuhn [1962] 1996, 3)

This statement points to the problem for the historiography of science of a- or unhistorical questions. That is, the very questions raised by historians of science should have a connection to the cultural and historical contexts of their sources, otherwise little progress is made in understanding. Without such a revised historiography of science and the questions it can productively pose, and sometimes even answer, the nature of the cuneiform sciences could never be realized in a fuller sense than that which remained focused only on its mathematical or strictly empirical aspects.

In the famous and oft-quoted postscript to *Structure*, Kuhn again underscored the perspectival nature of his historiography of science, saying, "Scientific knowledge, like language, is intrinsically the common property of a group or else nothing at all. To understand it we shall need to know the special characteristics of the groups that create and use it" (Kuhn [1962] 1996, 210). The common property of the cuneiform *eruditi* was in its essence encompassed by the parameters of *ṭupšarrūtu*. The group that stewarded this long and multifaceted tradition was that of the scribes and scholars discussed in this chapter. The scribes' craft (*ṭupšarrūtu*), its objects of inquiry, its knowledge domains and methodologies are the stuff of cuneiform science in its intellectual and textual culture.

The sociology of science and of scientific knowledge has not gone unrecognized in recent Assyriological scholarship. Since the later twentieth century, in addition to the continuation and expansion of textual analysis of astronomical tablets, research has indeed focused on contextual questions of social and intellectual dimensions (Rochberg-Halton 1991, Rochberg 1992, 1993, 2004, 2009a and 2009b, 2016a, and 2021; Ossendrijver 2011a and 2011b; 2021a and 2012b; Haubold, Steele, and Stevens 2019; Bowen and Rochberg 2020, subtitled "The Science in Its Contexts") not only concerning astronomical and astrological sources, but other areas of *ṭupšarrūtu* as well, such as medical, magical, and divinatory works. These works can provide a springboard for new research that raises useful questions for the ontological underpinnings of the epistemic culture of *ṭupšarrūtu*.

2

Cuneiform Texts and the Historiography of Science

I do not believe that there is any single approach to the history of science which could not be replaced by very different methods of attack; only trivialities permit but one interpretation.

– Otto Neugebauer, 1945

THE ENTRY OF CUNEIFORM TEXTS INTO THE HISTORY OF SCIENCE

The twenty-year period between Otto Neugebauer's two magna opera, *Astronomical Cuneiform Texts* (1955) and *A History of Ancient Mathematical Astronomy* (1975), saw the fruitful collaboration of Neugebauer and Abraham Sachs at Brown University. Together with Asger Aaboe at Yale (1955, 1964, 1980, 2001), they undertook further work on the mathematical structures, parameters, and methods of Babylonian astronomical texts (copied in Sachs 1955 [with Pinches and Strassmaier = LBAT]) in tabular and nontabular form, the latter including Diaries, Almanacs, Normal Star Almanacs, Eclipse Reports, and Observational texts, collectively referred to as nonmathematical astronomical texts (ADART). The project to publish the nonmathematical texts was completed by Hermann Hunger (Sachs and Hunger 1988, 1989, 1996a, 1996b; Hunger 1999, 2001 [with an appendix by Steele]), 2006, 2022 [ADART]). Neugebauer and Sachs's terminology of mathematical and nonmathematical astronomy became standard in the field, although inherent inconsistencies in these terms stem from the fact that a simple dichotomy cannot capture all the subtleties of the character of the various astronomical text genres and how they relate to one another.

A snapshot of the entire trajectory of the history of Babylonian astronomy from the heroic times of the founding Jesuit Fathers to the present is given by Ossendrijver (2012, 3–5; de Jong 2016). Accordingly, phase one ran from the discovery and decipherment of cuneiform astronomical texts in 1881 to 1935, when Neugebauer and B. L. van der Waerden entered the field, to the beginning of the most recent phase in 1990, when an understanding of cuneiform science as including omen divination and horoscopes alongside mathematics and astronomy came to the fore. In addition, the late twentieth century introduced social, cultural, and epistemological questions about the astronomical material with the goal of reconstructing not only the mathematical methods of Babylonian astronomy but its nature as a cultural phenomenon and, as well, the role, status, and place of its scribes in cuneiform intellectual history (Rochberg 2000, 2004; Ossendrijver 2011a and 2011b, 2021a and 2021b).

Continuing this development, in tandem with the direction the history of science as a whole has taken in the generation of scholarship since the 1960s and 1970s, it has become increasingly clear that not only divinatory but also magical, medical, and various procedural texts (Escobar 2017, 2019a, 2022, 2023) belong to the scope of the textual culture that was cuneiform science. Parallel and concurrent with the increasing acceptance of divinatory, magical, and astral medical texts into the framework of cuneiform sciences was another surge in the explication of mathematical astronomy by John P. Britton (in works from 1987 to 2010), Lis Brack-Bernsen (in works from 1969 to 2010), John M. Steele (works from 2000 to 2011), Noel M. Swerdlow (1998), and Mathieu Ossendrijver (2011–2012).

Certainly, one of the advantages for the cuneiformist interested in the history of science is the fact of the many extant and contemporaneous written sources stemming from the Old Babylonian period through the end of cuneiform writing itself in the late first century CE. This long-lived tradition is represented in scribal centers of the cuneiform heartland of Babylonia and Assyria, as well as to the west (see especially the divinatory texts from Emar in Rutz 2013; also in Rochberg-Halton 1988a, 30–35 and Koch-Westenholz 1995, 44–51), north (Hittite omens, see Güterbock 1987, 1988; Rochberg-Halton 1988a, 33–34; and Koch-Westenholz 1993), and east (Susa) of Babylonia. The bulk of the Nineveh collection of seventh-century BCE scholarly texts, generally referred to as the library of Assurbanipal, estimated in the many thousands of clay tablets (excluding writing boards and other media), is of the essence for the history of cuneiform science (Fincke 2004; Robson 2013; Finkel 2019). Tablets from Babylon and Sippar housed in the British Museum and the Vorderasiastisches Museum Berlin, as well as the tablets from Assur, the religious capital of the Assyrian Empire, are also relevant, as are Neo- and Late Babylonian-period texts from the Achaemenid and Hellenistic periods from the Seleucid Rēš Temple in Uruk to the sky god, Anu. Mathematical astronomical texts and horoscopic astrological texts belong to the Neo- and Late Babylonian periods.

Earlier collections shed light on the methods by which observed or observable phenomena functioned as signs and constituted a comprehensive divinatory system of knowledge. The knowledge from divinatory compendia included astral omens, extispicy – that is, the inspection of sheep's entrails – and other kinds of omens from physiognomy, animal and human behavior, dreams, and other categories of signs. In particular, astral divination and extispicy were the two scholarly disciplines favored by the Neo-Assyrian imperial state and then preserved, to a greater or lesser extent, within the Late Babylonian temples of Babylon and Uruk during the Achaemenid, Seleucid, and Arsacid periods (Reiner and Pingree 1975, 1981, 1998; Rochberg-Halton 1988a; Hunger 1992; van Soldt 1995; Koch-Westenholz 2000; Verderame 2002; Koch 2005; Heeßel 2021).

Medical diagnostic and therapeutic texts must also be included as part of a broad base of sources relevant to a study of knowledge in the cuneiform context, as discussed in Chapter 1 (Biggs 1990; Finkel 2000; Heeßel 2000, 2005, 2008; Scurlock and Andersen 2005; Scurlock 2006; Finkel and Geller 2007; Böck 2009, 2011, 2015; Geller 2010). Related thematically and in aim both to the medical and the divinatory repertoires, magical texts, including such text types as *namburbi*s (Maul 1994) and incantations, were integrally connected to the whole of scribal knowledge. Understanding the intersections of omen divination, magic, medicine, and the hermeneutics that methodologically join these disciplines to the lexical material, as Geller has shown (2022), and to each other will be key to writing a comprehensive history of cuneiform science as well as cuneiform intellectual culture in the future.

In essence, the totality of scribal learning, that is *kullat ṭupšarrūti* "everything pertaining to scribal knowledge" and *gimir ummânūti* "all scholarly expertise," becomes source material for the history of cuneiform science as seen from a more or less internal perspective. Within their ranks in either palace (seventh century BCE) or temple (fifth to third centuries BCE), the scribes engaged in a vigorous textual culture that produced a later legacy seen in the preservation of Babylonian astronomical tradition in foreign contexts, such as in the Levant, Egypt, Greece, Rome, India, and China.

The entry of cuneiform texts into the history of science from the modern historiographical point of view only began with astronomy because, being quantitative and rigorously predictive, those materials were found to be continuous with the later tradition of exact science and could be fitted to a somewhat modern model of scientific knowledge. At the time of cuneiform astronomy's rediscovery, the history of science was oriented toward an account of how modern science came to be, thus setting the tone for a mid-century acceptance of Babylon as its starting point. Later, however, the view of science in history came to encompass a wider set of epistemic values. The content and methods of cuneiform astronomical texts, then, could be viewed as part of the practices and purposes of scribal-scholarly learning as a whole. The resulting holistic approach to cuneiform intellectual historical sources put astronomy in a broad context with divinatory texts, horoscopes, astral medicine, and other kinds of divination (extispicy, morphoscopy, terrestrial omens, etc.) as parts of a system aimed at understanding the interconnection of human, divine, and world, both in its physical aspect and its textual representation. Defining cuneiform scholarly aims is critical for identifying the criteria for what science and knowledge were in the cuneiform tradition.

Despite the commonalities based on the focus on stars, planets, and the luminaries shared by astronomers and astronomical traditions cross-culturally, it is important to make more subtle distinctions that characterize any given tradition. One example, and there are many, is that of the difference between Babylonian astronomical science and Greek. If Ptolemy's *Almagest* can serve to represent the astronomical science of his day, the following in Book 1.1 underscores some significant differences from the Babylonian approach, despite some obvious points of similarity. Ptolemy said:

> We were drawn to the investigation of that part of theoretical philosophy [mathematics, because it is certain knowledge] ... especially to the theory concerning divine and heavenly things. For that alone is devoted to the investigation of the eternally unchanging.... this science, above all things, could make men see clearly; from the constancy, order, symmetry and calm which are associated with the divine, it makes its followers lovers of this divine beauty, accustoming them and reforming their natures, as it were, to a similar spiritual state. It is this love of the contemplation of the eternal and unchanging which we constantly strive to increase. (*Alm.* I.1, Toomer 1984, 36)

Ptolemy's invocation of the divinity of the heavenly bodies and the fact of their "eternally unchanging" nature, while sharing the interest in divine things, has a quite different orientation toward what makes for the divinity of the heavens and the heavenly bodies as compared with the Babylonian system.

Babylonian astronomy, too, was practiced within a tradition of knowledge heavily weighted in divination and a desire to read the "the heavenly writing," a metaphor for the celestial phenomena written by the gods upon the sky. This metaphor was used by Assyrian kings in the seventh century, but it became a lasting trope attested into the early modern period, if we think of Giovanni Gioviano Pontano's letters of a cosmic alphabet formed by the stars and planets (Grafton 1999, 6). But for Babylonian astronomers, who also saw the planets as the divinities, such as Marduk for Jupiter, Nabû for Mercury, Ishtar for Venus, the aspect of divinity in Ptolemy's conception as eternal and unchanging, or that moving images of this eternal divinity could be seen in the stars, was different from the idea of the gods as the authors of the future by writing on the heavens. As well, Ptolemy's statement that "the heaven is spherical in shape, and moves as a sphere; the earth too is sensibly spherical in shape, when taken as a whole; in position it lies in the middle of the heavens very much like its centre; in size and distance it has the ratio of a point to the sphere of the fixed stars; and it has no motion from place to place" (*Alm.* 1.2, Toomer 1984, 38) was not a Babylonian assumption. The arithmetic calculation of the recurrence of visible periodic phenomena by means of periods and period relations (Rochberg 2010a, 355–56) in Babylonian astronomy did not refer to the spherical world (see Chapter 7).

There were other considerations for the study of the heavens in cuneiform culture, even into the Seleucid period, as compared with those of Greek astronomical culture. The Babylonian objective in astronomy was not conceptualized in terms of the shape and motion of heavenly bodies, or of earth, much less considerations of sizes and distances of the luminaries with respect to earth. The importance of making such distinctions between historical sciences is to underscore the determinative role cultures play in the content of science and thus to provide a culturally informed historiography of science, for Babylonia or any other premodern science.

Twentieth-century scholars of Babylonian astronomy and astrology laid the necessary foundation for subsequent generations. However, integration within broader histories of science was slow in coming. On one hand, much of the work in mathematical astronomy was difficult for noncuneiformists or even other specialists in ancient astronomy, and on the other, that which was devoted to celestial omens, while not technical from a mathematical point of view, was equally difficult to integrate for the reason that ancient divination was not viewed by general historians of science as a part of their purview. Attention to the cuneiform scientific corpus, in all its dimensions, suffered from the fact that it was simply not accessible beyond the boundaries of the field of Assyriology.

General histories of science that told the story of science retrospectively from the present and projected contemporary scientific values upon past sciences misrepresented ancient cuneiform sciences, which were far too "other" for consideration. Presentist historiography, as it came to be called, came under especially heavy criticism by the historicist generation roughly between the 1960s and 1980s. Their critique took aim at Whig historiography, first defined by Herbert Butterfield, who said that Whig histories meant "to praise revolutions provided they have been successful, to emphasize certain principles of progress in the past and to produce a story which is the ratification if not the glorification of the present" (Butterfield 1931, 2). It was not so much a total repudiation of any relevance for the present, the inevitable and unchangeable vantage point of the historian, but, as he said:

> It is part and parcel of the whig interpretation of history that it studies the past with reference to the present; and though there may be a sense … in which it is inescapable, it has often been an obstruction to historical understanding because it has been taken to mean the study of the past with direct and perpetual reference to the present. (Butterfield 1931, 11)

Instead of "hunting for the present in the past," Butterfield stated the chief aim of the historian to be "the elucidation of the unlikeness between past and present" and the mediation "between other generations and our own" (Butterfield 1931, 10). This was a historiographical attitude that could have made a big impact on the study of Babylonian astronomy and cuneiform science, but in the 1930s it was too soon.

Whig histories continued to focus on and favor those aspects of a linear picture of scientific development pointing in the direction of modern sciences and to feature the episodes in the history of science that got something right. Results of this approach for premodern science were the omission, or sidelining, of significant episodes in the history of the sciences that did not join in the shaping of modern scientific thought. Such episodes were exemplified by the traditions of astrology, alchemy, divination, and magic. And so the related Babylonian sources would not be sought after for general histories of science.

Even where plainly some of the methods applied to the practice of ancient astrology, or magic, were compatible with science, gatekeeping seemed particularly rigorous when it came to "scientific thought," especially concerning material versus nonmaterial objects of inquiry, or material versus "occult" causes. Accordingly, there was no place for cuneiform science outside of the mathematical methods developed for astronomy, especially those found in Achaemenid and Seleucid Babylonian tablets, and the excellent results those produced for the mathematical prediction of, for example, lunar eclipses or planetary visibilities. That these scientific methods were used for astrological purposes and continued into the Hellenistic and medieval periods for the practice of horoscopy in Europe would only later and gradually come into focus.

What the latter-day critics of Whig histories of science found most problematic was, as Nicholas Jardine discussed (1986), that science in presentist terms was understood to be singular and absolute. The particular characteristics of scientific absolutism that Jardine critiqued (1986, 1) were especially its putative universal characteristics, including that it is characterized by a linear cumulative progress to truth. That model of the singular universal history of science met with methodological and historiographical challenges in large part because of the ahistorical nature of absolutism about science, and, consequently, it was discarded by any but naïve realists, of which few remain.

However, in the history of the cuneiform sciences, presentism remained a force in earlier literature, as when Late Babylonian mathematical astronomy, truly the first of the exact sciences in antiquity, was regarded as ideologically separate from other Babylonian corpora of knowledge, such as astrology and divination. Thus, said Neugebauer:

> No more drastic discontinuity in the history of ancient astronomy can be imagined than the creation of mathematical astronomy in the Babylonian ephemerides and procedure texts ... in the Seleucid period. If astronomical phenomena had been considered since the earliest Mesopotamian period as celestial omina ... the authors of the ACT material ... dropped all these traditional connections and analyzed lunar and planetary motion in a strictly mathematical fashion comparable only to the approach of Hipparchus and Ptolemy. (Neugebauer 1989, 392–93)

The confrontation of intellectual values between the astronomer-scribes who created or used mathematical tables and those who copied omens or earlier astronomical traditions, such as the series MUL.APIN (Hunger and Steele 2019), however, is not intrinsic to the sources, nor is the scenario that an alleged separate group of "mathematical astronomers" dropped traditional connections in order to establish their autonomy from the old omen tradition substantiated in texts. Indeed, in the Seleucid Greek environment of the ancient cities of Babylon and Uruk, temple scribes were dedicated to preserving textual traditions of diverse contents, such as omens, liturgy, magic, incantations, and medicine (Ossendrijver 2021a). The textual tradition of astronomy itself was maintained until the end of cuneiform writing, for the last dated astronomical tablet (Almanac W22340a, see Hunger and de Jong 2014) may have been written in the year 79/80 CE (Steele disputes this date, Steele 2019, 155).

More recently, historians of science have found it useful to engage in a different way with presentism (Jardine 2000; Tosh 2003; Harman and Métraux 2013). Laurent Loison defended a new presentism, defining this project of historical epistemology as "a way to solve the issue of presentism in the history of science" (2016, 29). As he put it:

> If one accepts that the objective of science is to produce true explanations, it seems difficult to dismiss all use of present knowledge to trace back and understand its history. This is why, as David Alvargonzález claims, "history of science is essentially whiggish" (Alvargonzález, 2013) – at least to some extent. The question is therefore no longer *if* we have to make room for presentism, but rather *how* we should use presentism. (2016, 29)

If, however, the relevance of presentism is predicated on accepting that the goal of science is "true explanations," it is not at all clear to me that such true explanations, or even explanation per se, can serve as the goal for all historical science, as McMullin already noted in 1984.

The question of what is more characteristic of science, its exclusive capacity to provide a true description of the world or its explanatory power, was a serious point of difference at that time (Ellis 1985, 1988). Instead, the goal of explanation seems to be a particular indicator only of particular forms of science produced in particular contexts. For astronomy in Babylonia, there are no explanations for the phenomena derivable from their mathematical methods. There are no explanations in any of the knowledge corpora of *ṭupšarrūtu* because explanation, in the form of a physical causal account, was not relevant in astronomy, astrology, or other kinds of omen texts. On the other hand, the question of explaining disease is a different case, and the scribes seemed genuinely interested in the etiology of illness.

METHODOLOGICAL PROBLEMS FOR A CUNEIFORM HISTORY OF SCIENCE

In an early paper titled "The History of Ancient Astronomy: Problems and Methods" (Neugebauer 1945), Otto Neugebauer pointed to the importance of chronological context for the identity of astronomy across the geographical regions of the greater ancient Middle East and Mediterranean from the third century BCE to late antiquity. Long before the terminology of globalized or localized knowledge entered academic discourse, Neugebauer was concerned with the effects of chronology, geography, and language upon the character of astronomical science as it transformed across the centuries from Babylonian to Greek, Greco-Roman, and European scientific cultures. As he said:

> Entirely different conditions underlie the astronomy in Egypt of the Middle and New Kingdoms than in the periods after the Persian conquest. Greek astronomy of Euclid's time has very little in common with Hipparchus' astronomy only a hundred and fifty years later. It is evident that it is of very little value to speak about a "Babylonian" astronomy regardless of period, origin, and scope. (Neugebauer 1945, 2)

Neugebauer's early attention to the dynamic between fixed geocultural contexts and the transformation of astronomical knowledge due to what he termed "transmission over time" set an agenda for much scholarship on the legacy of the Babylonian astronomical tradition. This approach was valid for other knowledge traditions as well, as more and more evidence for the legacy of the cuneiform scholars and *ṭupšarrūtu*, in all its diverse fields of specialization, came to light. The methods and problems of interest in this section, however, are different from those in Neugebauer's 1945 paper.

By problems, Neugebauer meant not only the difficulties in writing the history of ancient astronomy but also the technical problems that motivated ancient astronomers to develop certain methods or to answer certain questions. He was also concerned with the problem of the great lacunae in sources, the problem of explaining the rise of Babylonian mathematical astronomy, and why Greek astronomy took a leap forward immediately after contact with Babylonian knowledge. Other problems were those of the social and economic conditions that had an impact on intellectual culture and maintenance of learning, and yet another was chronology and historical change. By methods, he meant the methods devised by the ancients for solving astronomical problems, whether numerical or computational, mathematical, linear, or geometrical. At stake for Neugebauer was how astronomical questions were solved by such methods.

Rather than anything so systematic as astronomical methods themselves, what interests the chapters of Part 1 of this book is the historiographical problem we have in grasping the scientific culture of the cuneiform scribes, and the importance for historiography of the methodology best suited to answer questions, not only about historical forms of knowledge and their methodologies (mathematical, observational, etc.), but also about the ontological framework(s) for the particular phenomena of interest to historical sciences, cuneiform or otherwise. There is a strong perspectival dimension important to the history of science, which is to say we cannot simply assume that a given scientific culture was invested in natural knowledge or in an understanding of an external reality identified with nature. At least, for a responsible anthropology of science, this cannot be an a priori assumption.

A number of methodological problems attend the history of science in the cuneiform world. One is the lacunose nature of the abundant texts – and this pertains not only to the tablets themselves, which are frequently broken and fragmentary, but also to the great gaps in chronology and geography in between groups of sources. Astronomical texts are a case in point, as the corpus of ephemerides and procedures that formed the foundation of our knowledge of Babylonian astronomy cluster in the Seleucid period, with few texts bridging the gap to earlier periods, such as the earlier first-millennium compendium MUL.APIN (Hunger and Steele 2019). Scattered astronomical texts from the sixth to the fourth centuries analyzed most recently by John Britton (2008, 2009, 2010) represent intermediate stages before the fully developed systems of Babylonian mathematical astronomy emerged, but a continuous narrative for the history of astronomy in ancient Mesopotamia is far from attainable. If other traditions are taken into consideration, for example the divinatory texts, gaps in continuity are also the rule.

Finally, there is the historiographical question of the very nature of the investigation of the history of science in the ancient Middle East. That is, if the project is defined as a reconstruction of science in Babylonia and Assyria, what exactly is being reconstructed? The question is not new. Peter Dear's "What Is the History of Science the History Of?" comes to mind. As he said:

> Rather than studying the history of something that we always knew in advance how to identify, historians of science have turned more and more to studying how that "something" itself has been formed as a historical object, with no permanent,

transcendental identity. Everything thus becomes historicized and contingent, and the days of the history of science as an apologetic or celebratory enterprise serving present-day science have become, so to speak, a thing of the past. (Dear 2005, 390)

Unlike the sciences derived from the Classical tradition, it is an even more awkward kind of history where the very category for identifying what evidence is relevant or not is not an internal or native category at all, but an artifact of a different time and a different culture. In using a nonnative category, we must be conscious of our own filtration of the evidence and realize that, rather than aiming to discover in the cuneiform world that which we always knew to be science, we have the opportunity to allow cuneiform texts to suggest other ways in which knowledge was valued and phenomena were understood, and so to further expand our ideas about the nature of science historically.

Because there is no science, whether taken as knowledge, practice, or method, to bring back from cuneiform texts as something to which the scribe-scholars themselves were consciously committed, we cannot, strictly speaking, extract from these sources a narrative about what kind of, or how, science happened there from anything like "their own" point of view. Consequently, whatever we imagine for the cuneiform tradition must be congruent with its historical evidence, but what we deem relevant as evidence has clearly shifted over time as modernist demarcation programs that were invested in dichotomies of science/nonscience and the rational/irrational have been laid (almost entirely) to rest.

If we have given up our commitment to absolutes, facts, and a continuous teleological narrative for the history of science, with all its potential for anachronistic distortion of historical evidence, we need not give up trying to understand historical actors' interests in the phenomenal world. The interest on the part of cuneiform scribes in phenomena as signs, as events to be predicted, as symptoms to be described is the stuff of our copious source material extending across millennia of Babylonian scribal culture. If interest in phenomena – observable and imaginary – is the stuff of a newly historicized and redefined science, then we still have work to do for cuneiform texts of *ṭupšarrūtu*. There is every reason to approach this material with the conceptual tools we bring to the study of other scientific corpora, that is, observation, explanation, rationality, and prediction, if only to delimit their differences and attend to their particular orientation toward the phenomena defined in the texts themselves.

THE VIEW FROM OUTSIDE

The historiography of Babylonian astronomy not only deals within the framework of the cuneiform corpora of knowledge but also belongs to a wider "scientific *oikumene*" constructed of the ancient Middle East, the Western Mediterranean, Egypt, and India (see the works of Pingree from 1978 to 1998). Analytic and synthetic treatments of Babylonian scientific texts by cuneiformists are, therefore, of value for those outside the immediate field of cuneiform studies. Arne Hessenbruch's *Reader's Guide to the History of Science* took note of "Egypt and Mesopotamia" with the following statement:

> Mesopotamia and Egypt used to be poor relations in the history of science, useful only as a foil for the so-called Greek miracle, the birth of real science and mathematics. However, new archaeological discoveries, the study of newly-uncovered texts, and reappraisals of those already known from the point of view of an enlarged range of questions have changed this image in the last few decades. More than in other areas of the history of science, older syntheses and the popular expositions based on them may prove to be at best obsolete, or even misleading. (Hessenbruch 2000, 196)

By the turn of the millennium, the field had indeed entered a new phase, one changed by some new sources but principally by a new historiographical sensibility, alluded to in the previous section. But ancient Middle Eastern knowledge corpora, certainly no longer a foil for their Greek counterparts, still present us with problems of understanding, because that mass of textual source material does not speak in our customary language of science, that is, of the kind of knowledge *scientia* or *episteme* are supposed to represent, nor of nature or a clearly defined category of the physical.

In the same *Reader's Guide to the History of Science*, under "Greece: General Works," the position of ancient Middle Eastern science vis-à-vis Greek is mentioned again. There, Elmer Yglesias (2000, 311–12) pointed out that the period just after World War II saw a marked increase in the attention given to ancient Greek science, and soon thereafter, according to the general outlines of the entry, cuneiform, that is, Babylonian astronomical science was introduced to the field by Neugebauer, whose *Exact Sciences in Antiquity* of 1969 Yglesias deemed "narrowly internalistic" (2000, 311). Neugebauer's *Exact Sciences* was indeed the closest reading of ancient scientific sources from the Middle East and ancient Mediterranean made available at that time. It provided a detailed look at the "internal" content of ancient science, but narrow it was not.

From the late 1960s and early 1970s to the early 1990s, Geoffrey Lloyd established a new foundation for the study of Greek science with a series of books from *Polarity and Analogy: Two Types of Argumentation in Early Greek Thought* to *Demystifying Mentalities* and *Problems and Methods in Greek Science* (1966–91). With reference to Lloyd's work, Yglesias pointed to the progressive engagement of the history of science with anthropology, philosophy, and social history, that is to say, not the internalist work of those interested in Babylonian astronomy. Lloyd's attention to methodological and historiographical issues set an intellectual stage for the special issue of *Isis* in 1992 titled "The Cultures of Ancient Science," with contributions by Pingree, Lloyd, Martin Bernal, Heinrich von Staden, and Rochberg. This collection of papers was something of a breakout piece, in which the ancient Near East was given a hearing in the premier journal of the American History of Science Society alongside ancient Greece.

There, von Staden offered a key methodological critique for the very approach to Greek science still valuable today for ancient science in general, which he termed the problem of "affinities and elisions." He showed that historians of ancient science privileged Greek science over that of other Mediterranean and Near Eastern cultures at their own risk of distortion, particularly when it resulted in the claim to the invention of science and the scientific method. These claims, he argued, would not have sat well with the ancient Greeks themselves, whose scientific writers show an awareness of and interest in, and were influenced by, what Arnaldo Momigliano famously termed "alien wisdom," most notably, but not only, that of Egyptians and Chaldeans (Momigliano 1990). Von Staden's paper, however, was not aimed only at Hellenocentrists. As he said, "'Hellenocentrism' is but one manifestation" of more fundamental problems that "arise from the oft-tacit interpretive deployment of 'affinity,' 'continuity,' and 'origin' – notions conditioned in part by the history of the modern reception of classical antiquity" (1992, 582). These problems, we can say now in looking back, are the consequence of a Whiggish history of science geared to accounting ultimately for the origins, roots, and rise of modern science. Pingree developed his argument against this position, calling it "Hellenophilia," providing, as he put it, "an apologia for my claim to be a historian of science with as serious and thorough a purpose as are the topics that we usually find discussed in history of science classrooms or in the pages of *Isis*" (1992a, 554).

While affinities, continuities, and origins were highlighted for writing the history of science retrospectively from the modern perspective, acknowledgment of the essential asymmetry and difference between the modern and the premodern eras began to create a gap, an intelligibility gap – what Ernest Gellner called the Big Ditch – between two "worlds." That is to say, the intelligibility gap produced a plurality of invalid world systems of traditional cultures before the Ditch but a single true world system of science after (Gellner 1979, 145–47).[1]

Yglesias took the 1992 *Isis* symposium as indicative of a changing perspective, where Greece would be viewed in the larger context of the whole of the ancient Mediterranean, including the ancient Middle East, meaning Babylonia and Egypt. "The Cultures of Ancient Science," as the issue was called, did point toward an emergent pluralism, as well as an enlarged framework for thinking about science in antiquity that has become increasingly normative since that time. China too, already a subject for the history of science in its own right, took on added importance, as Lloyd in works spanning 1996–2006 and in his collaboration with Nathan Sivin (Lloyd and Sivin 2003) showed, in reaching out beyond the boundaries of Hellas for a broader understanding of the nature of ancient science as a whole.

It is not only that the history of ancient science has come to recognize a wider geographical framework for its evidence, one that sets both the cuneiform and the Classical and Hellenistic Greek sciences within a wider cultural ambit, indeed within the framework of West and East writ large, but that the progressively historicized view of what science knows, does, and is has also changed the way we identify, read, and understand ancient texts relevant to what we think of as the history of science.

DEBATES IN THE HISTORIOGRAPHY OF SCIENCE

It has been nearly fifty years since Hayden White described historiography as "hedged about with taboos" and charged historians with failing "to recognize that every historical discourse contains within it a full-blown, if only implicit, philosophy of history" (White 1978, 126–27). In a dramatic indictment for his time, White saw that "the presence in a historical work of a formal theory used to explicate the relationship between facts and concepts is enough to earn them the charge of having defected to the despised sociology or of having lapsed into the nefarious philosophy of history" (White 1978, 126). By now, however, the recognition that all

[1] He called it, "one colossal, or should I say, abysmal, chasm" (146); and see also Geller 1982, 188–200.

historical scholarship rests upon certain philosophical ideas, themselves colored by culture and historical moment, determining how a historian analyzes and interprets her subject, has become methodological canon. The historian need not always lay bare the underpinnings of her historiographical method, but, as Lloyd said, "It is better to be self-conscious about it" (Lloyd 1992, 565).

I think conscious awareness of one's historiographical method and its underpinnings is essential when the historian's subject is science and no less so when the subject is the science, or sciences, of the cuneiform world of circa 2000 BCE to (at least) 100 CE, with its long legacy into the sixteenth century. Preceding me in this project is Eleanor Robson, whose *Ancient Knowledge Networks* (2019) focused on the modern (and ancient) construction of the institutional framework for cuneiform knowledge. Her deconstruction of our terms "library" and "tradition" reflects a social-historical and sociology of science approach with which I am in sympathy. Here, however, I focus not on the institutional level or the aspect of scholarly networks but on the way Babylonian knowledge reflected the aims of its practitioners.

To that end, an issue that somehow keeps surfacing regarding ancient science and its pursuit of knowledge about signs, or astrology, or magic should be addressed, and that is whether the twentieth-century problem of the demarcation of science is relevant for our approach to these subjects. Delineating the cuneiform sciences of astronomy and astrology is not definable in the terms of Karl Popper's famed demarcation project, although early to mid twentieth-century historiographies frequently separated Babylonian astronomy from astrology along demarcationist lines (discussed in Chapter 3). As is clear in the volume of papers edited by Massimo Pigliucci and Maarten Boudry (2013), the persistent issue for demarcation has settled on how to decide on the line separating science from pseudoscience (Pigliucci 2013, 9), and this serves as the central point of discussion throughout the volume. But as Popper defined the project in his 1935 (reissued in 2002) *Logic of Scientific Discovery*, it was not about pseudoscience. In his words, "the problem of finding a criterion which would enable us to distinguish between the empirical sciences on the one hand, and mathematics and logic as well as 'metaphysical' systems on the other, I call the problem of demarcation" (Popper [1935] 2002, 11). Pseudoscience per se was not the issue, but demarcation addressed a broader question in the theory of knowledge at its most fundamental level. In this way, Popper challenged the Vienna Circle positivists' commitment to induction as a method of generating reliable knowledge from observation.

Before the goal of demarcation settled on the project of distinguishing science from pseudoscience (which term does not occur anywhere in the text of *The Logic of Scientific Discovery*), all the nonsciences were targeted, including metaphysics, and many of the ideologies of Popper's day (famously, psychoanalysis and Marxism; see Popper 1963). Such questions of epistemology and meaning (according to the positivists, only empirical statements were granted meaning, metaphysical statements not) have little to no bearing on our present historical investigations, as the mode in which cuneiform scribes sought meaning in phenomena was distinct from natural science. Nevertheless, although the science/pseudoscience demarcation is anachronistic in the discussion of Babylonian astrology and celestial divination, the tendency to make such a division persisted, as discussed further in Chapter 3.

Demarcation became a matter of science versus pseudoscience only after the middle of the nineteenth century, when the term science took on its more or less familiar sense and the term pseudoscience was first used. Pseudoscience involves deliberate imitations of science, couching unscientific claims in such a way as to pass for science and gain credibility. In the cuneiform context, no evidence has ever emerged for such credibility contests, nor for an attempt on the part of any group to imitate some system of knowledge in order to establish authority. In historical contexts prior to the nineteenth century, and extending into the ancient world of cuneiform texts, attributing pseudoscience to such a world is what Nick Jardine would call "vicious anachronism" (2000, 253, and throughout). The fact that there are legitimate historical reasons for viewing ancient and medieval astrology, magic, or alchemy as sciences is not a weapon against modern scientific standards but simply shows the historicity of the methods, goals, and questions considered amenable to the sciences of the past.

An example of the employment of such ahistorical criteria for evaluating ancient and medieval sciences comes from nearly 100 years ago from the staunch positivist George Sarton. He called astrology "a very remarkable scientific system," saying, "a history of science which did not include an account *of the purer aspects* [my emphasis] of astrology and alchemy would be incomplete" (Sarton 1924, 85). Sarton's defense of these two ancient sciences comes in the context of his otherwise harsh critique of Lynn Thorndike's *A History of Magic and Experimental Science during the First Thirteen Centuries of Our Era* (1923), which included much within its purview that Sarton considered unscientific and worse, for example, divination and magic. Sarton's judgment exemplified the demarcationist use of epistemic standards extraneous and foreign to the historical evidence in question – textbook vicious anachronism. In this regard, Sarton's long demarcationist shadow was cast upon the journal *Isis*, founded by him in 1912 as the History of Science Society's official publication, until finally in 2002 the term pseudosciences disappeared from its annual bibliography.

The historiographical premise that science is constant in all its aspects across the ages, besides underpinning attempts at demarcation outside of the modern context, leads to other problems for the history of science. Apart from the demarcationist program, the history of science was conscripted into a critique of scientific realism, but this too required an assumption that the goal of science was constant across the ages, namely, to discover the real and true workings of nature. That antirealist argument, known as the pessimistic (meta)induction (Laudan [1981] 1984), begins with the idea that the sciences of the past were largely wrong, or based on insufficient knowledge, and thus deficient in some way when compared with more recent sciences. According to this position, past sciences engaged with phenomena, some of which turned out not to refer to anything (aether and phlogiston are commonly cited), and consequently such sciences were based upon mistaken ontologies, inherently untrue.

Despite the appearance of a grossly ahistorical measurement of premodern achievements by modern standards in all this, the purpose of the pessimistic (meta)induction was not to impugn the history of science but to claim that if the history of science shows us the error of our ways, we can suspect that our current scientific theories will also prove false in time, thus weakening the argument for scientific realism. The image of the history of science as a "graveyard of dead scientific theories and abandoned theoretical posits" was disputed by Moti Mizrahi (Mizrahi 2016, 263–65), who found it a tactic based on cherry-picked cases to argue against realism in science. He called the graveyard metaphor a "philosopher's myth" (Mizrahi 2016, 265).

The pessimistic induction engaged in a direct comparison of historical evidence for science against modern and erased the differences between them that relate to different sorts of representational (or other) goals of science. The graveyard metaphor was originally Peter Lipton's, who said:

> The history of science is a graveyard of theories that were empirically successful for a time, but are now known to be false, and of theoretical entities – the

crystalline spheres, phlogiston, caloric, the ether and their ilk – that we now know do not exist. Science does not have a good track record for truth, and this provides the basis for a simple empirical generalization. Put crudely, all past theories have turned out to be false, therefore it is probable that all present and future theories will be false as well. That is the pessimistic induction. (Lipton 2005, 1265)

The pessimistic induction thus validated the writing of the history of science retrospectively and emphasized continuities. That the truth of theories in science is temporary, regardless of some of their long-lived successes (Ptolemaic astronomy, humoral theory of disease, Newtonian physics), does little for the history of science, but it was not meant to contribute to a greater understanding of historical sciences. Indeed, the history of science when viewed as a graveyard of methods, ideas, theoretical posits, and ontological claims, even if we are keen to see the connective tissue of scientific methods and ideas over time, places emphasis on modern assessments of how they failed rather than what might have accounted for their success.

Mizrahi disputed the graveyard metaphor on different grounds. He reckoned that there are fewer "abandoned theoretical posits" (e.g., celestial spheres, luminiferous aether, stellar influence, humors, phlogiston) than those that are still accepted (e.g., enzymes, atoms). The history of science, however, has more to offer than comparisons to modern science, or evidence for answering the question of whether science itself displays a continuous progressive development over time. The existence of continuities, of which there are especially many when it comes to methods (rational, empirical, logical, mathematical, etc.), does not preclude all the differences that matter if our goal is the understanding of science in its particular historical time and place. Georges Canguilhem put it succinctly when he said, "history is not an inverted image of scientific progress" (in Delaporte 2000, 45).

Regardless of the important methodological continuities over time and across cultures, room must be made for the parts that do not map onto the expected parameters of science from a modern perspective. The parts that do not map are of the essence for coming to terms with distant sciences. These are the phenomena that construct world-versions – the humors in the Greek theory of disease or the celestial spheres of Greek astronomy, the *dynameis* (δυνάμεις "influences" or "powers") of the stars in Greek astrology – ontologies implied in the terms and descriptions contained in historical sources. For a science that was once as valid and central to scientific communities as Greco-Roman and medieval astrology, what makes it of interest to the history of science is not whether its principal theoretical posit – stellar influence – was wrong or right but how it shows us something of critical importance to the way its practitioners understood the world.

From a cultural standpoint, the history of science affords us insight into the different theoretical posits of our sources and, consequently, the opportunity for the historian to discern different worlds assumed and *made* by scientific cultures of the past. In light of attested cultural differences, the pessimistic induction does not make room for those differences in the theoretical frameworks of the past. The pessimistic induction, which assesses the truth or falsity, not the success or failure, of scientific theory only works if we level the ontological playing field from premodern to modern science.

In spite of the vast difference in worlds and their ontologies from antiquity to modernity, it is both possible and useful to show the lines of continuity as well as discontinuity in the problems, goals, and methods of sciences over time. The continuities reassure us that the ancient astronomical sciences, for example, operated within a world of celestial phenomena that we recognize (planetary and lunar synodic phenomena, in particular) and came to certain results from effective predictive methods that we immediately recognize as kindred to our own methods of scientific inquiry. The discontinuities require historians, through the evidence of dissimilarity and difference, to confront a world as it was conceived long ago, even millennia ago in the case of the cuneiform world. Babylonian divination, as an example, operated within a conceptual framework in which phenomena did not function as parts of a unified physical universe with definable physical laws. Thus, it does not seem scientific to us on the ontological level in spite of certain logical or systematic characteristics it displays in an epistemological sense.

LOCAL KNOWLEDGE AND PLURALISM ABOUT SCIENCE

A change in the status of ancient science for historians was the result of broader intellectual trends – the so-called interpretive, social, and linguistic turns; the work of new historicists, contextualists, and poststructuralists; the sociology of scientific knowledge, constructivism, local knowledge, and science as practice – concerns about presentism, and, generally speaking, a drift toward pluralism. The anthropology of science belongs squarely within this complex of ideas and approaches.

For the historian of premodern and non-Western sciences, the most important turn was the turn away from both a universal conception of science as transcendent of history and culture and the assumption that such a universal science is instantiated by Western science. Thomas Kuhn looms large in this process. In *Structure of Scientific Revolutions* ([1962] 1996), he was explicitly committed to granting other people their "worlds" (see further in Chapters 4 and 5). This shift in focus to difference and discontinuity in the content and values of knowledge over time and across geographical boundaries raised new questions for the interpretation of ancient science and insisted that it be understood on its own terms, not simply as a precursor and certainly not an immature version of later science.[2] Clifford Geertz deserves mention in this context as well, as his thoroughgoing pluralism and focus on the local mirrored the change in the history of science. Referring, in his characteristically vivid terms, to the aim of science, he said:

> The notion that the surer grasp of unshapely and incongruent, even unique, particulars is as proper an aim of science as the abstractive formulation of exceptionless regularities – and is, often enough, more illuminating as well – has grown steadily more acceptable over the last quarter century [referring to the period 1975–2000] as rationalism stumbled, positivism evaporated, and the "prism face of Newton" (the image is Wordsworth's) faded from view. The notion that all knowledge aspires to the condition of mathematical physics ... lacks the air of simple obviousness that it had even a few short years ago. Everything, from the philosophical reconsideration of the nature of natural law to the spread of perspectival, observer-dependent explanation, has strengthened the claims of case-based knowledge to scientific standing. (Geertz 2000, x)

2 When I say "on its own terms," I will no doubt be taken to subscribe to the overused emic perspective. In a sense I do, but not in the usual way, where emic is opposed to etic because it is "internal" or meaningful to the insider, rather than "external" or meaningful to the outsider. Boyarin 2009, 9 note 8, says, "These terms [emic and etic], produced out of the structural linguistic opposition between 'phonemic' and 'phonetic' distinctions, bear no analogy whatsoever to internal and external or subjective and objective. Phonemic distinctions are distinctions that make a difference of meaning within a given language, while phonetic distinctions are simply distinctions in sound that make no difference within that language ... Phonemic distinctions are observable from the 'outside'; indeed one of the first tasks of a classical structural description of a language was for the researcher to make, through observation, a list of the phonemes of the language. As such they are just as objective and repeatable as a list of phonetic differences, and they are the only important categories for describing that language." I too see "emic" distinctions as meaningful to insiders but discernible to outsiders, and essential to our ability to make analyses and interpret ideas from other cultures. Actors' categories are accessible to the anthropologist and historian alike, or, as Collins (2008, 101) put it, "insofar as analysts are going to develop categories of their own – analysts' categories – to do the work of explanation, those categories will have to be built upon actors' categories."

He was referring here to the nature of modern science, but the implications for its history are clear.

One of the major effects of the change in the historiography of science overall is that the aim and focus of the history of ancient science is no longer to trace the ultimate origins and roots of modern science and thus to support the notion of scientific unity and reductionism to the one unified truth about nature's laws, properties, and kinds. Cuneiform astronomical texts, in their role as directly antecedent to Western astronomy, may once have earned their place in the history of science precisely on the basis of this intimate and significant connection between ancient and modern. But the singling out of astronomy, particularly mathematical astronomy, as *the* Babylonian contribution to the history of science did its share in reinforcing the historiography of science as something to be written backward from modernity to antiquity so as to support the model of unity.

Late Babylonian astronomy became a powerful symbol of what the ancients were able to accomplish and seemed on the face of it to testify to the fact that Babylonian scientific aims were consistent with ours: to obtain knowledge and predictive control of phenomena and to discover their autonomous "laws." Because the ancients were attempting to describe or predict the same phenomena as did later astronomers, that is, the periodic synodic appearances of planets, the first visibility of the moon, and so on, the question of what it was the ancients observed and perceived about these natural objects, and why, was not, in the modernist and presentist historiographical mode, raised as a historical problem.

The value of Babylonian astronomy to the history of science, once tied to its status as the earliest chapter in a story of the development of Western planetary theory, is now, in addition to that, that it can show how much more complex were the relations among various aspects of the cuneiform intellectual culture. Late Babylonian astronomy can no longer be viewed as an isolated corpus of scientific mathematical ephemerides, more or less easily translatable to Western counterparts. The Babylonian theory of the planets, as Swerdlow pointed out in the introduction to his 1998 book of the same name, had its cultural place in a world of diviners and temple benefices. The multiple epistemic values of Babylonian astronomical science, therefore, especially those not obviously congruent with later astronomical science, are valuable sources for the anthropology of science.

Presentism, with its commitment to ahistorical scientific universalism, has mostly given way to a postmodern emphasis on so-called local knowledge, meaning the various culturally embedded systems of knowledge of all kinds, science included. This strategy to historicize, to challenge absolute standards or criteria, and to recognize pluralities is certainly not new. Kuhn ([1962] 1996) already voiced the need to render even nature in accordance with historical context and continued to develop that position further (Conant and Haugeland 2000, 216–23). In that vein, Bruno Latour would later proclaim, manifesto-like: "From cultural relativism we move on to 'natural' relativism" (1993, 106). Latour worked this subject persuasively in his provocative *We Have Never Been Modern* (1993) and seems to have been the first to employ the plural "natures" to represent the multiple conceptions of the interaction of human beings and the phenomena derived from anthropological and historical approaches to science. But the idea was in the air, as we see in a statement of Joseph Rouse from two years prior that questioned why "narratives of modernity" in the history and philosophy of science were so difficult to root out and overcome. There, Rouse went so far as to suggest that "the idea that there is a 'natural world' for natural science to be about, entirely distinct from the ways human beings as knowers and agents interact with it, must ... be abandoned" (1991, 162). The idea that cultures are relative but only science, and with it nature itself, is universal meets a number of challenges on the basis of historical texts of the premodern era, and I would count the cuneiform texts as primary among them.

Countering the picture of an ahistorical canon of science, local knowledge in a variety of historical and social contexts has been in the foreground now for two generations of historians. In part, this was a consequence of the very shift in aim from determining the teleological arc of how our present scientific state of the art came to be, to raising questions as to the nature of knowledge in former periods without an overemphasis on how they evolved or did not evolve to the present day. While in some circles the unity of science together with its reductionism has given way to an emphasis on disunity, localism, and pluralism, in other circles the controversial unity model, particularly in the context of natural kinds and ontological monism is still a valid question (Tahko 2021).

In the philosophy of science, unity, whether in the form of laws (Tahko's "structure of theory," in Tahko 2021, 10) or kinds (Tahko's "structure of reality," Tahko 2021, 10; 4 figure 1) has been considered anew. A position has been articulated in which the fact of historical and contemporary epistemic pluralism can be integrated with ontological monism (Tahko 2021, 21–28, esp. 23). Sandra Mitchell asked, "If science is representing and explaining the structure of the *one* world, why is there such a diversity of representations and explanations in some domains?" (Mitchell 2002, 55; emphasis in the original). She ruled out what she called competitive pluralism, as a symptom of "the ultimate aim of science ... to resolve the conflicts by adopting the single unfalsified, or 'true,' or overwhelmingly supported winner of the competition" (Mitchell 2002, 56). In this case, she explained, pluralism is a temporary and eliminable condition, where unity is still the goal. Such varieties of competitive pluralisms, in her view, do not account for all kinds of pluralism, and she defined another kind, namely "compatible pluralisms," to cover legitimate alternative explanations of phenomena. Mitchell's context of analysis was the biology of social insects. As she put it, "what in fact explains the existence and characteristics of division of labor in social insects, what appeared at first sight to be the 'same' phenomenon requiring a single explanation, will be itself contingent on the particular features and pathways that occur in each case" (Mitchell 2002, 67). This integrative pluralism, in which science "makes room for compatibility" (Mitchell 2002, 68), however, preserves ontological unity; indeed, it was never meant to disavow it (Tahko 2021, 24).

On the other hand, historians of science, whether they are concerned with *episteme*, *scientia*, or *ṭupšarrūtu*, similarly deal in epistemic pluralism, that is, the "diversity of representations and explanations in some domains" referred to by Mitchell. However, epistemic pluralism, as a way to account for the particular ways diverse historical sciences construed the objects of their knowledge and the central reasons for knowing them, invariably runs into the question of ontological pluralism unless differences are viewed as the superficial symptom of divergent worldviews, not the fundamental difference in and therefore pluralism of worlds.

The consequence for our approach to cuneiform disciplines of *ṭupšarrūtu* is that the measure of their value for the history of science is not dependent upon how they compare or do not compare to later sciences, or how we can assimilate or accommodate their views within a supposed unity of science. Although the absence of a conception of nature means that the cuneiform cultural perspective on the physical world and its phenomena had fundamental differences from that of later Western science, the kinship between these traditions, particularly in the astral sciences, enlarges our view of the scientific enterprise and underscores the historiographical value of pluralism.

THE ORDER OF NATURE AND THE OBJECT(S) OF SCIENTIFIC KNOWLEDGE

The question of the relationship of cuneiform knowledge to the history of science must confront the historiographical problem posed by the idea of nature. This is so because of the identification of scientific knowledge, by and large, with natural knowledge. Scientific ideas in premodern non-Western contexts, however, flatly contradict this identification, notably within cuneiform culture (Rochberg 2016a). This observation can become an issue regarding the unity or disunity of science, both from an epistemological and/or an ontological standpoint. Cuneiform and other non-Western sciences are replete with evidence of the investigation of phenomena that share any number of features with later natural sciences, such as astronomy, astrology, and medicine, but these were not oriented to or shaped by questions about nature. Rather, they addressed questions of interest and importance to their practitioners for other reasons and other motivations than the investigation of physical properties within a materialist or physicalist system. Ontological reductionism to the material world of physics, therefore, has built in limitations for premodern non-Western sciences.

How the scribes classified or divided the phenomena of their known world into kinds, that is relevant cultural kinds rather than natural kinds (Rochberg 2016a, 95–101; 2019) is a difficult ontological question. We have hints as to such classes or divisions of things in the use of determinatives in the cuneiform script, but those only go so far. In any case, we must resist the temptation to resort to Plato's analogy in the *Phaedrus* (265e) that the world comes to us with readymade classes of things to be divided, as he said, "at the joints," usually taken to mean those determined by nature.

Without a reference to nature – and therefore no sense of it comparable to that held by Greek, medieval, or early modern European natural philosophers, certainly not by modern Western scientists – our understanding of the particular orientation of cuneiform scholars to the phenomena that interested them needs to establish itself on different grounds. Our construal of cuneiform bodies of knowledge and their associated activities, as well as every aspect of the philosophy of such knowledge – epistemology, reasoning, causality, observation, explanation, and prediction – must answer, within the limits of our own conceptual limitations to understanding their framework, to the ancient native conceptual grounds, the universe of cuneiform observation, interpretation, and theorization.

On this basis, neither the act of knowing nor the world as an object of experience or imagination can be defined a priori. Both are rooted in a particular perspective, a point of view and orientation to the world.

Both nature's definition and its attachment to science, especially to modern science, have been at the foundation of the history of science since its inception, when the goal of the field was to formulate an account of the origins and rise of modern science. Historians of science, however, cannot afford to take the nature of nature as self-evident. What we make of the natural world is a direct function of our historical and cultural ways of knowing and our linguistic ways of describing. In the cuneiform world, as in any number of premodern and non-Western cultures, engagement with phenomena, physical and otherwise, did not give rise to a term or a conception of nature as an independent objective reality. While it is true that a separate domain, equivalent to what we call nature, did not have its own word in the perceptual, conceptual, or linguistic world of the cuneiform scribes, a variety of texts reflect various ways of systematically looking at what *we* call natural phenomena in that cuneiform world.

If indeed the compendia of ominous signs, medical therapies and incantations, pharmacological ingredients (the compilation of medicinal plants URUANNA = *maštakal* is as yet unpublished, but for secondary literature, see Böck 2011, 2015), plant and stone lists (Schuster-Brandis 2008; Stadhouders 2011, 2012), as well as astronomical texts, both mathematical and nonmathematical, and later astrological and astromedical works belong to science in Assyria and Babylonia, the objective of all that knowledge can neither be defined in terms of nor reduced to a desire to know or explain nature. Nor can the practices associated with these handbooks and compendia be assimilated to the various objectives of later natural sciences, or even natural philosophy, namely, to understand the order of nature, nature's laws and causes, or the relationship between nature, man, and God. These are not epistemic motives or scientific values that fit the picture created by the written corpora of *ṭupšarrūtu*, nor of its practices, whether associated with divination at the Neo-Assyrian court, the nightly watch of the heavens in Babylon from the eighth to the first centuries BCE (ADART), or the complex tradition of medical diagnosis and treatment, fully integrated with the use of magic.

In sum, with regard to premodern sciences not derived from the Classical tradition, fundamental assumptions about science as a form of thought, a method of inquiry, a set of practices, or indeed an inquiry whose object is nature and natural phenomena construed in a particular (Western) way are ripe for reassessment. For the premodern and non-

Western context, a historicized, pluralized, and relativized view of science is key to taking account of other cultures' ways of thinking about and describing the phenomena of interest to them, be they heavenly bodies, sheep entrails, plants, stones, or the behavior of animals, demons, or ghosts.

In his recent defense of pluralism in the history and philosophy of science, Hasok Chang (2012) rejected both monism and reductionism in scientific theory on the basis of the utter complexity of nature and our attempts to understand it. He offered that the monistic view of nature was but a stage, if a necessary stage, in the development of science.[3] He said:

> It is possible that science needed to be launched on the strength of hubris, which made the whole enterprise seem both doable and worth doing: that we *could* grasp *the* truth about nature! Perhaps early scientists needed to believe that nature was fundamentally simple enough for them to be able to understand it. Newton had God on his side, and proceeded with the faith that there was one truth about God's creation and with the grace of God he could find it – how else would anyone muster the conviction that one simple equation could cover all the universe? (Chang 2012, 255; emphasis in the original)

Thus it makes sense as one looks at the progress of science since the Scientific Revolution, using the term here as a convenient chronological marker, to question the utility and even validity of such monism and/or reductionism.

For the cuneiform world, though, we are in even less familiar terrain, both before nature *and* before God – or perhaps better expressed as without nature and without God. What seems unavoidable is that whereas no assumption seems more fundamental to our conception of science than nature, if we are truly to embrace a pluralism about historical sciences, we ought to move beyond epistemic pluralism to consider the possibility of an ontological pluralism in the evidence from non-Western premodern contexts (as discussed in the chapters of Part II).

In order to get inside the culture of cuneiform science and knowledge so as to define what inquiry seems to have been about in Babylonian and Assyrian texts, what may once have been considered epiphenomenal – things like the use of divination as a legitimizing tool of the state, the scope

[3] Cf. Sandra Mitchell for a similar but earlier view that complexity warrants pluralism in science, and that pluralism is not simply a strategy on the road to the unity of science. That is, that "history shows us that many sciences do not exhibit a diminution in the multiplicity of theories, models, and explanations they generate. This 'fact' of pluralism, on the face of it, seems to be correlated not with maturity of the discipline, but with the complexity of the subject matter ... Pluralism reflects complexity" (Mitchell 2002, 55).

of the Late Babylonian scribes' expertise in texts from incantations to ephemerides, the continuous preservation of millennia-old traditions of omens, not to mention reconstructing familial connections and networks among the scribes of Uruk or of Babylon – should be taken into ever greater account as we further explore *ṭupšarrūtu* in the contexts in which it makes sense to speak of culturally embedded scientific knowledge. Today, we accept that many kinds of science have existed, instantiated by evidence from many periods and cultures of history, including those of the ancient Middle East. A productive approach to this cultural diversity and multiplicity is an anthropological one that pays specific attention to ways of worldmaking, that is, how scientific communities engaged with the phenomena of the world of *their* making. Such an approach was not in the forefront of the historiographical approach to our subject adopted among its twentieth-century interpreters, although it was slowly emergent from the 1990s onward. It is to the foundational work of three representatives of the twentieth-century historiography of ancient science that I now turn.

3

The Early Shaping of a Modern Historiography of Ancient Astronomical Sciences

This chapter brings an analysis of the historiographical underpinnings of three twentieth-century historians of science. In particular, I am interested in how they reflected or challenged contemporary philosophical presuppositions about science. This chapter is not meant to offer a survey of twentieth-century historians of astronomical sciences, but only to fix a lens on three notable historians, Otto Neugebauer, Olaf Pedersen, and David Pingree, in order to exemplify the philosophical foundations for their historiographies and to illustrate the point that there is an inextricable relation of philosophy to historiography. Equally as important, such analysis provides a basis for reckoning with our own approach to the long-lived intellectual traditions of the Babylonian scribes in astronomy and related scribal scholarly disciplines in both Assyria and Babylonia and extending into realms of the Hellenistic period beyond the homeland of the Tigris and Euphrates region.

OTTO NEUGEBAUER AND OLAF PEDERSEN

Two important modern histories of ancient mathematical sciences still considered authoritative today are Otto Neugebauer's *The Exact Sciences in Antiquity* (1969) and Olaf Pedersen's *Early Physics and Astronomy* ([1974] 1993, 2nd ed.). Neugebauer's *Exact Sciences in Antiquity* was first published in *Acta historica scientiarum naturalium et medicinalium*, volume 9, E. Munksgaard, 1951; then with Princeton University Press, 1952; Brown University Press, 1957, 2nd edition; and then reprinted by Dover Publications in 1969. I will refer to the 1969 edition throughout this chapter. According to John North's obituary of Pedersen, *Early Physics and Astronomy* "was a revised translation of the first volume of what was to have been a three-volume history of physical science by Pedersen and Mogens Pihl. The 1974 edition carried both their names, but the second edition was published in the name of Pedersen alone" (1998, 214).[1]

Because of the similarity of the chronological and cultural range of these two books, the differences in their aims, and their wide distribution, the works of Neugebauer and Pedersen serve to exemplify the historiographical treatment of the ancient exact, or mathematical, sciences in the middle/late twentieth century. These books will therefore provide the primary focus for the following discussion, but where it may be relevant to shed further light on Neugebauer's point of view, I will also adduce his *A History of Ancient Mathematical Astronomy* (Springer, 1975). That work had not appeared in time for Pedersen to have known about it when writing *Early Physics and Astronomy* ([1974] 1993).

Olaf Pedersen (1920–97) was a younger contemporary of Otto Neugebauer (1899–1990). They were both in Copenhagen during the 1930s, although not at the same time. Neugebauer escaped Nazi Germany and taught Babylonian mathematics and astronomy at the Mathematical Institute from 1934 through 1939 (Brack-Bernsen 2016, 113). Pedersen arrived the year of Neugebauer's departure and earned his degree in theoretical physics at the Niels Bohr Institute, then called the Institute of Theoretical Physics (1939–43; Hoskin 1998, 5.33). As Pedersen did not turn to the history of science until the 1950s, the question of influence upon the young Pedersen by Neugebauer is improbable.[2]

Following his period of sanctuary in Copenhagen, Neugebauer would go on, in 1939, to take up a professorship in the History of Mathematics Department at Brown University, while Pedersen went to Aarhus, where he founded the Department for the History of Science at Aarhus University (North 1998, 212). That department became an independent unit in 1965. Pedersen was also involved in the creation of the Danish journal *Centaurus: International Magazine of the History of Mathematics, Science, and Technology* in 1950, for which he served as

[1] I thank Helge Kragh of Aarhus University for drawing my attention to John North's obituary of OP, as he was known to his colleagues and friends.

[2] I gratefully acknowledge the help of Professors Helge Kragh and Kirsti Andersen with the biographical details of Olaf Pedersen's student years in Copenhagen.

one of the editors alongside Kirsti Andersen, Ole Knudsen, and Kurt Møller Pedersen. According to Victor Thoren in a review of *Centaurus*:

> There is some imbalance toward the exact sciences, reflecting the facts that the physical sciences have tended to dominate the history of science until fairly recently and that Denmark has a long tradition of research in the history of mathematics. And for this reason the articles have tended to be somewhat positivistic – although already in Volume 3 (1953) there appeared a special collection titled "Essays on the Social History of Science." (Thoren 1991, 304)

Pedersen's interest in the history of science was, therefore, part of a later stage in his career. Upon receipt of his PhD, Pedersen taught physics in high school.[3] As John North pointed out, "His view of the subject [history of science] was formed by his experience in teaching physicists in terms that they could not only understand but value. The general history that he wrote, *Early Physics and Astronomy* (1974), gives a very good idea of his ambition to spread the good word" (North 1998, 213).

While Neugebauer's *Exact Sciences* and Pedersen's *Early Physics and Astronomy* both stemmed from the authors' involvement with the exact sciences of mathematics, physics, and astronomy, the books evince clear-cut differences in their respective presentations of science in antiquity. Despite these differences, which will be detailed in what follows, it is also evident that embedded in each lies a shared image of science as the form, practice, and method of knowledge that best represents – or has the capacity to represent – the way the world of physical phenomena works. Science, according to this wholly modern image, is made distinctive as a form of knowledge by reason of its object of inquiry, the universe of physical phenomena. The definition of science to which both authors subscribed is further distinguished by the fact that a quantitative methodology is held to be key to the practice of science through the analogy of nature as a physical mechanical system with its own eternal ineluctable laws, best described in the language of mathematics.

This image of science, or one very like it, lies at the root of what has come to be known, after Karl Popper (*Logic of Scientific Discovery*, [1935] 2002, 11), as the problem of demarcation (discussed in Chapter 2). A long-standing debate ensued in the philosophy of science centered on finding workable criteria for establishing the special epistemic warrant for scientific knowledge. Thomas Nickles (2013, 105) called the philosophical demarcation of science "a conservative exercise in

[3] Helge Kragh, personal communication.

exclusion," as it aimed (and still aims) to dismiss the pseudo- and non-sciences. The aim of demarcation has thus largely been of particular relevance for modern and contemporary sciences, as well it should, but it will have different ramifications in historical contexts. The fact that science is now widely considered not to have constituted a thoroughgoing unity throughout its historical development, but to have been the product of a diverse range of interests, values, and questions emerging within local communities, is a complicating factor for the demarcation project. For the history of science, even the history of the exact, or mathematical, sciences, there can be no reduction to a single set of criteria for separating real science from other forms of thought; the historian's aim, if nothing else, is to recognize and come to terms with that view.

Although the debate was a central concern for philosophers of science of the middle of the twentieth century, by 1983 Larry Laudan proclaimed "the demise of the demarcation problem" due to its being "intractable" (Laudan 1983, 125) and in 1996 called it "an unqualified failure" (Laudan 1996, 86). By "failure," he meant that "there is apparently no epistemic feature or set of such features which all and only the 'sciences' exhibit" (Laudan 1996, 86). As a result, Laudan argued, demarcation ceased to be a viable philosophical issue.

The issue, however, is far from dead and in fact has seen a recent resurgence (Mahner 2007, 2013; Hansson 2013; Ladyman 2013; Pigliucci 2013), though it must be admitted that the recent discussion is not concerned particularly with the history of science. Regardless, some of what was at stake for the demarcation problem in the mid twentieth century was endemic to the histories of the exact sciences in antiquity of both Otto Neugebauer and Olaf Pedersen. As such, it seems worthwhile to examine how demarcation criteria are made manifest in their historiographies.

Both Neugebauer and Pedersen give an account of the development of the exact science of astronomy from Middle Eastern antiquity to the Middle Ages. Both deal with the ancient Middle East and its relation to Greece, but to opposing ends. Pedersen saw the progress of science as a matter of the triumph of knowledge over belief, of science as the systematic knowledge of nature purified of practical or religious motives, and he saw science as having made this decisive transformation for the first time in Greek civilization. In contradistinction to this position vis-à-vis the relation between the ancient Middle East and Greece, Neugebauer (1969, 1) viewed any overemphasis on the philosophical content of Greek science as a distraction from the essential component of exact science, which was its mathematical content and quantitative methods in all their technical detail, and the role played in the progress of science by what he called accurate knowledge.

Science, in Neugebauer's estimation, was signaled by the construction and application of mathematical models to prediction in astronomy, producing accurate knowledge of the phenomena. As such Neugebauer saw science as an achievement of the ancient Middle East, specifically of the Babylonian astronomers. Moreover, how mathematics and astronomy were developed in Babylonia and transmitted from there west to Greece and east to India became clear indicators for Neugebauer of the contact and relations between ancient as well as medieval Indian and European civilizations and of the translatability (or commensurability, although he would not have used either of those terms) of the exact sciences across cultures and history.

Each author gave different degrees of weight to various subjects in accordance with the goals of his account of the nature and progress of exact science. Neugebauer focused exclusively on the mathematical methods of Egyptian, Babylonian, and Hellenistic Greek astronomy, while Pedersen admitted a greater role for philosophy in scientific knowledge, particularly in terms of the understanding of physical properties such as matter and motion in the cosmos.

As stated in his preface ([1974] 1993, ix), Pedersen sacrificed "a detailed account of Egyptian and Babylonian science before the Greeks," citing "the number of introductory works on these topics which have recently been presented by eminent scholars" as the reason for his decision not to offer his own account of them.[4] He must certainly have been referring to Neugebauer's *Exact Sciences in Antiquity* as well as, perhaps, B. L. van der Waerden's *Science Awakening* (vol. 1 on mathematics, 1st ed. 1954, 5th ed. 1988; vol. 2 on astronomy, 1974). At the same time, however, Pedersen's view of what the history of science meant, that is, the emphasis on philosophical rationality and an understanding of physical law, is consistent with the decision to omit most of the evidence for astronomical knowledge in Babylonia and Egypt.

[4] Michael Hoskin (1998, 5.34), in his obituary of Pedersen, said in reference to Pedersen's *Early Physics and Astronomy*, "The last quarter-century has seen a remarkable development in history of science, and today the cursory treatment of Babylon and Egypt, and of the Islamic Middle Ages, would be seen as a major defect."

In sharp contrast to Pedersen's understanding of the history of science, Neugebauer viewed the philosophical commitments of his historical subjects as irrelevant to the content of their quantitative methods. Neugebauer also resisted the idea of producing a synthetic history of science on the basis of "philosophical constructions." Accordingly, he wrote:

I do not consider it as the goal of historical writing to condense the complexity of historical processes into some kind of "digest" or "synthesis." On the contrary, I see the main purpose of historical studies in the unfolding of the stupendous wealth of phenomena which are connected with any phase of human history and thus to counteract the natural tendency toward oversimplification and philosophical constructions which are the faithful companions of ignorance. (1969, 208)

Not only did Neugebauer deem "philosophical constructions" detrimental to the historiography of science, but he also disparaged the impact of philosophy on the actual development of science.[5] About claims to the influence of Plato, for example, Neugebauer said:

I think that it is evident that Plato's role [in the development of mathematics] has been widely exaggerated. His own direct contributions to mathematical knowledge were obviously nil. That, for a short while, mathematicians of the rank of Eudoxus belonged to his circle is no proof of Plato's influence on mathematical research. The exceedingly elementary character of the examples of mathematical procedures quoted by Plato and Aristotle give no support to the hypothesis that Theaetetus or Eudoxus had anything to learn from Plato. The often adopted notion that Plato "directed" research fortunately is not borne out by the facts. His advice to the astronomers to replace observations by speculation would have destroyed one of the most important contributions of the Greeks to the exact sciences. Plato's doctrines undoubtedly have had great influence upon the modern interpretation of Greek sciences. But if modern scholars had devoted as much attention to Galen or Ptolemy as they did to Plato and his followers, they would have come to quite different results and they would not have invented the myth about the remarkable quality of the

[5] The year following the publication of *The Exact Sciences in Antiquity*, Douglas McKie of the University of London delivered his inaugural address titled "Science and History" (McKie 1958). In Joseph Agassi's account of this address, he relates how McKie dealt with the relationship of history to philosophy, saying, "A historian of science ... may be personally interested in the philosophy of science ... *qua* historian, however, he has no use for it; his business is simply to study the rise of scientific ideas as it took place" (Agassi 1963, 6). Agassi, whose historiographical critique is aimed at what he judged to be Baconian inductivists, further remarked that "the demand that the history of science be separated from the philosophy of science is a modification of Bacon's demand that science and metaphysics – and thus the history of science and the history of metaphysics – be separated. He based this demand on his claim that science begins with observation and metaphysics with speculation" (Agassi 1963, 83). Of course, Agassi fell on the postpositivist side of the historiographical debate.

so-called Greek mind to develop scientific theories without resorting to experiments or empirical tests. (Neugebauer 1969, 152)

This idea was reiterated in Neugebauer's later history of mathematical astronomy as follows: "I see no need for considering Greek philosophy as an early stage in the development of science . . . [its effects were] disastrous enough to forestall better insight for centuries" (Neugebauer 1975, 572). Because Neugebauer saw the history of astronomy as a matter of development through transmission and a continuity from the ancient Middle East to the later West (and East, meaning India), he repudiated any significance for aspects of cultural institutions not governed by the epistemic values of empiricism and quantitative methodology.

Unlike Neugebauer, Pedersen saw astronomy as integrally bound up with natural philosophical aims to understand the physical cosmos, and that tradition originated within Greek philosophy. In his chapter on medieval astronomy, Pedersen commented that "astronomy was also interesting to philosophers, partly as material for epistemological discussions on the nature of scientific hypotheses and theories, partly as a basis for metaphysical disquisitions on the ultimate structure of being, dependent, in turn, on a supreme cause, just as the motion of the planetary spheres was caused by the *primum mobile* outside the fixed stars" (Pedersen [1974] 1993, 214).

In the middle of the twentieth century, the general assumption that physics was the model science no doubt helped both Neugebauer's and Pedersen's histories of the exact sciences to find a receptive audience in a field of the history of science, where early science was, and still is, almost by definition on the periphery of more critically relevant developments perceived as beginning in the early modern period. The position that physics is the science that reveals the true mechanistic laws of nature and was itself original to Greek antiquity, was nowhere stated explicitly by either author. It was perhaps revealed implicitly, however, in the treatment of mathematics as the transcultural and transhistorical language of science. Thus the exact, or mathematical, sciences are almost by definition also transcultural and transhistorical in the sense that they are readily transmitted across cultures and over long stretches of time.

To identify Neugebauer's position on the nature of astronomy and its relation to the history of science, we can look at the conclusion to *The Exact Sciences in Antiquity*, where he said:

We have come to the end of our discussion, which has brought us back to the civilization of the Middle Ages from which our journey started. By patiently following the connections of mathematical and astronomical theory we moved from period to period and from civilization to civilization. Our road often went parallel to the road pointed out by historians of art, religion, alchemy, and many other fields. This is not surprising. It only underlines the intrinsic unity of human culture. The role of astronomy is perhaps unique only in so far as it carried in its slow but steady progress the roots for the most decisive development in human history, the creation of the modern exact sciences. (Neugebauer 1969, 177)

This statement conveys the view that the history of astronomy, and with it the history of exact sciences more broadly, comprises a continuity from Babylonian antiquity to the beginning of the modern era (the eighteenth century). Indeed, in the later work, he drew a clear and unbroken line of development in mathematical astronomy "beginning in the latest period of Mesopotamian civilization and ending with the discovery of the elliptical orbit of Mars, thus preparing the way for Newton's dynamics which brought to its conclusion all ancient mathematical astronomy" (Neugebauer 1975, 1). Already in the *Exact Sciences*, Neugebauer included "Mesopotamian civilization" in the history of exact science for its having produced science in the form of mathematical astronomy but specifically excluded Egypt:

In summary, from the almost three millennia of Egyptian writing, the only texts which have come down to us and deal with a numerical prediction of astronomical phenomena belong to the Hellenistic or Roman Period. None of the earlier astronomical documents contains mathematical elements; they are crude observational schemes, partly religious, partly practical in purpose. Ancient science was the product of a very few men; and these few happened not to be Egyptians. (Neugebauer 1969, 91)

Pedersen, on the other hand, saw a discontinuity in the history of the development of modern science between Greek and as he put it "pre-Greek" civilizations, by which he meant Babylonia. He stated:

There are many reasons in support of the view that the achievements of the Greeks were the most decisive factors in the origin of modern science. It is true that pre-Greek civilizations had made important findings both in mathematics and technology, and that the Babylonians, in particular, had succeeded in giving astronomy a concise mathematical form contemporary with the unfolding of the geometrical astronomy of Eudoxos, Apollonios, and Hipparchos. Nevertheless, the briefest of comparisons between Babylonian and Greek science shows the extent to which the Greeks succeeded in reshaping and developing scientific methods by a conscious effort reflected in their philosophy of science. The elegance and perspicacity of the Babylonians are undeniable and admirable. But to a modern scientist it is equally impossible not to feel more at home in the worlds of Eudoxus or Archimedes. All later science is indebted to the Greeks because of their adherence to the rational concept of natural law. Their discovery of how suitable hypothetical-deductive theories are to the description of nature, their introduction of the concept of mathematical models, and their first tentative uses of the experimental method are features of such lasting importance that the development of science would have taken a completely different course without them. (Pedersen [1974] 1993, 150)

Each author saw as an element relevant for the history of science the relation of ancient to modern science. As key to its identity as science, ancient historical science, each agreed, should be comparable or relatable to certain criteria still held by modern science, and yet those criteria by which the ancient sources would be judged were viewed differently by each author. Whereas Neugebauer had no problem in seeing an evolution leading to modern astronomy beginning with Babylonian mathematical astronomy, by virtue of its mathematical language and its capacity to produce what he called accurate knowledge, Pedersen understood the history of science as continuous not with Babylonian efforts but only first with Greek methods and the impact on science of Greek philosophy, in particular the distinct form of rationality conceived in Greek philosophy.

An important difference between Neugebauer and Pedersen can be seen in terms of how far to extend the line of continuity between modern and ancient science. Neugebauer saw the origins of the exact sciences firmly rooted in Babylonian mathematical astronomy. Pedersen ([1974] 1993, 20), in the style of Gomperz, Schrödinger, and Burnet, placed the origins of science with the Greeks, seeing Classical Greece of the fifth century as no less than a "scientific awakening" in which "nature began its divorce from myth." No matter how sophisticated the use of the astronomical units, parameters, period relations, and observations of the Babylonian astronomers was, Pedersen remained convinced that "the briefest of comparisons between Babylonian and Greek science shows the extent to which the Greeks succeeded in reshaping and developing scientific methods by a conscious effort reflected in their philosophy of science" (Pedersen [1974] 1993, 150).

The radical split between the historiographical outlines of Neugebauer (1969, 1975) and Pedersen ([1974] 1993) is at odds, however, with the fact that beneath the difference between them on how to conceive of the origins of the exact sciences lies a deeper shared commitment to the idea of a continuous development of science from antiquity to modernity and the related view that it was possible to demarcate science from nonscience, superstition, and dogmatic belief. This becomes clearer when probing their respective attitudes toward astrology.

Neugebauer's and Pedersen's agreement as to the distinctive identity of science by virtue of its being amenable to mathematical description and prediction (Neugebauer) or its logical and rational nature (Pedersen) also comes across in their respective attitudes toward astrology. Despite Neugebauer's defense of the importance of astrology to the historian, astronomy and astrology in his as well as in Pedersen's history come to be seen as separable along modern lines of demarcation that adjudicated the boundaries between science/nonscience, rational/irrational, and progressive/regressive in accordance with what was then a central position in the philosophy of science. This aspect of the historiographies of Neugebauer and of Pedersen echoes the effort in the philosophy of science of the early to mid twentieth century to demarcate those very boundaries in such a way as to maintain the integrity and unity of science throughout its history.

Whereas the mainstream of the history of science focused on the mathematical sciences, another stream, begun in the late nineteenth century, provided a new context within which to consider the relationship between ancient astrology and astronomy. This came to the fore as the result of the scholarly movement, contemporary with the period of the recovery and decipherment of the cuneiform astronomical texts, which would ultimately contribute significantly to the understanding of and comparison between the astral sciences of the ancient Mediterranean West and the ancient Middle East. It was a comprehensive project to collect and edit ancient Greek astrological manuscripts (CCAG), undertaken by the Belgian classical philologist and historian Franz Cumont, together with classical philologists Franz Boll and Wilhelm Kroll (see Chapter 1). Wouter Hanegraaff (2012, 314) pointed to the turn of the century as the productive period not only for the CCAG (Cumont, Boll, and Kroll 1898–1953), but also for Berthelot and Ruelle's *Collection des anciens alchemistes grecs* (1887–89). This alternative stream would later (1944–54) include André-Jean Festugière's *La révélation d'Hermès Trismégiste*.

But the reception of astrology, magic, and alchemy was not without its detractors. In 1924, the then-recognized dean of the history of science in the United States and founder of the American History of Science Society, George Sarton, in his review of the fourteen-volume magnum opus of Lynn Thorndike, *A History of Magic and Experimental Science* (1923–58) consigned the contents of most of it to the other side of the moon as far as his own definition of science was concerned. His diatribe was filled with dramatic flair, calling the history of science "a history of victories and conquests; permanent conquests over the chaos; a growing light eating up the darkness. No other human activity lends itself to such a heroic and stupendous tale" (Sarton 1924, 83). The cumulative, progressive, rational, and skeptical science was all that magic and divination was not.

A generation later, even Neugebauer saw the value of astrological texts principally as conveyors of astronomical knowledge. He made this statement in the oft-quoted one-page rebuttal of another one of George Sarton's condemnations, this time of Sarton's review in *Isis* volume 41 (1950, 374) of Margaret S. Drower's publication of the Mandaean *Book of the Zodiac* (Drower 1949), which Sarton dismissed as "a wretched collection of omens, debased astrology and miscellaneous nonsense." Neugebauer's defense came out in *Isis* volume 42 (1951, 111) under the ironic title "The Study of Wretched Subjects."[6]

The project to differentiate science from other forms of knowledge, particularly astrology, alchemy, or magic, as Sarton did, is a direct consequence of the demarcationist aims of logical positivism. It is interesting in the present context to note the role astrology played in Popper's own rejection of the positivists' focus on inductive methods as key to demarcating science. He said:

> Astrologers, more especially, have always claimed that their "science" was based upon a great wealth of inductive material. This claim is, perhaps, unfounded; but I have never heard of an attempt to discredit astrology by a critical investigation of

[6] Joseph Agassi saw the dichotomy differently. He said, "I very much sympathize with Thorndike's demand that the fact that almost all astronomers before Newton were astrologers should not be concealed; but I see no reason to transcribe all the details of these astrological views. I also sympathize with his view that 'magic and experimental science have been connected in their development,' even though I disagree with him on this point; one can hardly speak of the development proper of 'magic,' (i.e., magic, mythology, superstition, etc.) and science has evolved *out of* 'magic' not *together with* it" (1963, 11; emphasis in the original). He added, "Undoubtedly, Thorndike's greatest fault, and the one which renders his work so useless and infuriating, is that he puts all errors on a par, in a true Baconian fashion. The worst mediaeval superstitions are lumped together with Kepler's animism: all errors are prejudices and superstitions" (1963, 11). In view of the critique Agassi levels at what he calls the "inductivist history of science" – that it distorts the history of science by considering as part of its development only those ideas that have proved correct or that are based on sound principles and derived from observation of facts (inductive method) – it is puzzling to me that at the same time he seemed to be saying that those ideas (such as astrological influence) that are the outcome of hypothesis alone fall on the wrong side of history.

its alleged inductive material. Nevertheless, astrology was rejected by modern science because it did not fit accepted theories and methods. Thus there clearly was a need for a different criterion of demarcation. (Popper [1963] 2004, 345)

Popper saw demarcation as the central problem for the philosophy of science, and his own dominance in the philosophy of science discourse of the mid twentieth century surely had an impact on how science was treated by historians as well.

Pedersen set astrology in the context of a mythopoeic and divinatory worldview where a personalized view of phenomena gave rise to the idea of the gods' power over nature and with it the use of magic to prevail upon the gods' will. While he understood the development of astronomy as the only reasonable outcome of a society's desire to manage its world, its calendar, or its intellectual curiosity about the heavens, he found astrology "difficult" to reconcile with what he saw as a natural inclination toward the development of physical science. He offered that if asked how to explain the phenomena of the universe, the "pre-Greek" ancients would reply in mythological terms. He prefaced his claim to the mythopoeic attitude with the following comment: "It is difficult to believe that the various pre-Greek civilizations could have been completely devoid of any intellectual structure equivalent to the theoretical science of later times, for it is impossible to visualize a people who never pondered on the origin and development of the material world" (Pedersen [1974] 1993, 7). Pedersen visualized the ancient Egyptians and Babylonians as solving their questions about the world by imagining that "individual natural phenomena are thought to be caused directly by the free will of the God presiding over that part of nature" (Pedersen [1974] 1993, 8). Furthermore, as he wrote:

In consequence of this conception of nature the ancient world also practiced many different forms of divination, that is the art of divining the hidden purposes of the gods by means of portents in the visible world. . . . A particular case was astrology, or divination by the stars, which spread in the last centuries before our era all over the Mediterranean countries from Mesopotamian where the Sun, Moon and planets were worshipped as the most powerful god. (Pedersen [1974] 1993, 9)

Pedersen maintained a clear separation between astrology and astronomy because of what he viewed as the religious overtones of astrological thought, that is, its dependence upon undemonstrable "beliefs" rather than demonstrable mathematical "theories." As he said of Ptolemy's astrological work, "a special historical role was played by his *Tetrabiblos* in which the principles of astrology were described in a secular manner without mythological or religious overtones, as if this particular form of divination was a kind of applied astronomy" (Pedersen [1974] 1993, 76–77). The insertion of the word "secular" served to classify Ptolemy's *Tetrabiblos* on the side of science, that is, as referring to strictly mechanical relationships between and among the planets and luminaries, unsullied by "belief."

Finally, in reference to what he called the victory of astrology in medieval Europe due to anxieties about the Black Death in the middle of the fourteenth century, Pedersen revealed his true feelings about the practice of astrology when he said that "many sober-minded philosophers and scientists tried to dam this stream." He cited in particular a fourteenth-century work of Nicole Oresme in which the "combination of mathematical insight and an argument of probability undermined one of the fundamental assumptions of astrology, but had no practical consequences." He concluded that "the old superstition was irresistible and the Renaissance period greatly enforced it, even reducing the *Almagest* of Ptolemy to a mere introduction to Albumasar and other astrology authors" (Pedersen [1974] 1993, 215–16).

Neugebauer also mounted a defense of Greek astrology as science because of its mechanistic view of the cosmos:

> To a modern scientist, an ancient astrological treatise appears as mere nonsense. But we should not forget that we must evaluate such doctrines against the contemporary background. To Greek philosophers and astronomers, the universe was a well-defined structure of directly related bodies. The concept of predictable influence between these bodies is in principle not at all different from any modern mechanistic theory. And it stands in sharpest contrast to the ideas of either arbitrary rulership of deities or of the possibility of influencing events by magical operations. Compared with the background of religion, magic and mysticism, the fundamental doctrines of astrology are pure science. (Neugebauer 1969, 171)

Consistent with his position in "The Study of Wretched Subjects," we can still see the effects of the demarcation problem in the background. The passage concludes by saying, "Of course, the boundaries between rational science and loose speculation were rapidly obliterated and astrological lore did not stem – but rather promoted – superstition and magical practices. The case of such a transformation from science to humbug is not difficult to exemplify in our modern world" (Neugebauer 1969, 171). Neugebauer assessed astrology to be a conservative practice, a methodological hodgepodge, as is clear in a statement from the later work:

> The claim of astrology that it is based on age-old wisdom or on revelation by a deity could only reduce the incentive for a search for better astronomical methods. Consequently we see that astrological practice is often based on

antiquated methods, on a conglomerate of diverse, even contradictory, parameters – facts most welcome for the analysis by the historian but very detrimental for the progress of astronomy. (Neugebauer 1975, 5)

Neugebauer's evaluation of astrology in relation to science saw the importance of context as a means to compare it against what he deemed even less rational practices, such as religion, magic, and mysticism. Astrology could be science when construed as a mechanistic system, but this was only the case for Greek astrology, where stellar influence through planetary rays was argued by Ptolemy (*Tetrabiblos* 1.24, Robbins 1971, 115 and 280–81 note 1) and the structure of Greek physical theory provided a framework for mechanization. David Pingree defined it this way:

Astrology grew out of a union of aspects of advanced Babylonian celestial divination with Aristotelian physics and Hellenistic astronomy. . . . The product was the supreme attempt made in antiquity to create in a rigorous form a causal model of the *kosmos*, one in which the eternally repeating rotations of the celestial bodies, together with their varying but periodically recurring interrelationships, produce all changes in the sublunary world of the four elements. . . . In other words, ancient Greek astrology in its strictest interpretation was the most comprehensive scientific theory of antiquity, providing through the application of the mathematical models appropriate to it predictions of all changes that take place in a world of cause and effect; it is not surprising, then, that it was called simply *mathēsis*, or "science" by Firmicus Maternus and others. (Pingree 1992, 560)

Similarly in Neugebauer's account, Greek astrology was a science because it displayed mechanistic causal features, such as that of the direct physical effect of the planetary rays upon earth.

The idea that after 500 BCE, on the evidence of horoscopes, Babylonian astrology had departed from divination by celestial omens to become a mechanistic – naturalistic – system has no basis. The laconic enumerations of planetary positions in cuneiform horoscopes do not indicate that the Babylonian astronomers had given up their view that divinities were involved in the functioning and meaning of the heavens and its signs because the textual culture that supported celestial divination (and extispicy) was still in place, and no indication from outside horoscope texts exists to testify to a discrepant position vis-à-vis the gods. A horoscope depicted the situation of the heavens at the time of a birth to be taken as a composite sign from the heavenly gods for what lay in store for a person's life. The definition of astrology as a system in which the planets and their mutual relationships are determinative of change in the sublunary realm by the action of physics and the geometrical nature of planetary positions, while true of Ptolemy's *Tetrabiblos*, does not aptly characterize Babylonian astrology. A further historiographical implication of turning Babylonian astrology into a purely physical science is potentially what Peter Harrison called the "science causes secularization" narrative, which he opposes (Harrison 2017, 47). My understanding of the cuneiform horoscopes leads me to reject the "science causes secularization" narrative as well in this instance.

With reference to Babylonian astrology's relation to astronomy, Neugebauer was more concerned to revise previous assumptions, taken largely from ancient Greek authors – for example, the first-century BCE Diodorus Siculus' reporting of "Chaldean" astral science (in *Bibliotheca historica* Book II.29–31; Oldfather 1998, 445–57; Jones and Steele 2018, 335–41 for a new edition of the entire passage) – as to the "mystical" or "astrological" character of Babylonian astronomy, particularly that of the late Babylonian period (after 500 BCE). He credited Epping and Kugler's pioneering work in Babylonian mathematical astronomy with setting the foundation for a revision of the attribution of mysticism or astrological motivation for Babylonian mathematical astronomy (Neugebauer 1969, 97).

There were in fact ancient concerns with demarcating astronomy and astronomical knowledge from physical theory and whether astrology was a science, as discussed by Alan Bowen (2007). Bowen identified the crux of this problem as the result of the interest of Greek astronomers of the second century BCE in the Babylonian horoscopy. He said: "The Hellenistic reception of Babylonian horoscopic astrology gave rise to the question of what the planets really do and whether astrology is a science. And this question in turn became one of defining the Greco-Latin science of astronomy, a project that took Aristotle's views as a starting point" (Bowen 2007, 330). He further pointed out that Geminus, the first-century BCE author of the treatise *Introduction to the Phenomena* (*Introductio astronomiae*; Evans and Berggren 2006), demarcated astronomy from physical theory and subordinated the planetary models that accounted for the planetary stations and retrograde arcs to the philosophical arguments for the uniformity of planetary circular motion, arguments grounded in the perfection and divinity of the heavenly bodies such as one sees in Plato's *Timaeus* (Bury 1929) and Aristotle's *De Caelo* (Guthrie 1971).

Therefore, for Geminus, astronomy could only account for *apparent* irregularity, while physical theory accounted for *real* regularity. Roughly two to three centuries after Geminus, Ptolemy proposed that astronomy "is a surer guide to reality than what the Stoics call physical theory or, in his terms, theological and physical philosophy" (Bowen 2007, 352), and finally, that "Ptolemy's elevation of scientific philosophy, specifically astronomy, in cognitive status entails that the scientist's arguments for key propositions previously taken as belonging to physical theory will by themselves be sufficient to establish the truth of these propositions and they will in fact be the best kind of argument possible" (Bowen 2007, 352). Although this ancient demarcation problem is distinct from that of the twentieth century, each had a strong epistemological motivation, and there are clear reverberations between them. In fact, some aspects of the ancient debate concerning what is real, and therefore "true," would contribute to the foundation upon which was based the modern philosophical debate around real and true versus pseudo- and false science.

At the appearance of Babylonian horoscopic methods to derive data for the moon and planets – whether by observational (meaning from Diary texts or goal-year methods, Brack-Bernsen 1999; Brack-Bernsen and Hunger 2008; Gray and Steele 2008; Gray and Steele 2009; Steele 2011b) or computational means (meaning Babylonian mathematical methods and ephemeris tables) – Greek astronomers were confronted by the problem of understanding planetary synodic appearances, all of which were included in Babylonian astronomy (first visibilities, first and second stationary points, and last visibilities for all five naked-eye planets, Mercury, Venus, Mars, Jupiter, and Saturn). Only after the second century BCE did these phenomena have to be taken into account within the terms of Greek cosmology with its commitment to geocentric uniform circular motion of the divine heavenly bodies. The ancient demarcation problem came out of the concern of Greek physical theorists as to the nature of astronomical, that is, quantitative, explanation versus that of physical theory.

Returning to the historiographical context within which to understand both Neugebauer's *Exact Sciences* and Pedersen's *Early Physics and Astronomy*, the dominant historiography of science in the mid century was clearly influenced by logical positivism and its heirs and, consequently, was burdened by the demarcation problem. Both the history and the philosophy of science would soon come up against a more socially informed approach, where the kinds of problems to which historical sciences sought answers were no longer analyzed by historians solely in terms of true and universal knowledge of nature, but rather in terms of the social, political, and cultural values characteristic of the historical context of a given source.

Spearheaded by Kuhn's *Structure of Scientific Revolutions*, which shifted focus to the questions of what prompts change in science, or what Kuhn termed normal science, and how, in fact, science could give rise to different world perspectives, the historical and social contexts within which science functions took center stage and gave rise to new questions about the nature, image, and purpose of science over time. This was radically different from previous nods to the importance of the relationship of science to society. Indeed, as mentioned before (p. 82), there had been an issue of the journal *Centaurus* devoted to historical aspects of the impact of science on society. In his introduction to this volume, L. Rosenfeld said, "The very decision to create a special Commission for the study of the social relations of science in their historical perspective demonstrates the importance attached by scientists and historians to this particular aspect of the history of science" (1953, 7). Although Rosenfeld took pains to argue for the legitimacy of inquiry into the history of science's relations to society, his argument was aimed at the relevance of this historical research for understanding modern science's relation to and impact on modern society. Furthermore, the historical model according to which science had a single evolutionary line of development was assumed, just as it was assumed that society (meaning complex societies continuous with the modern European West) had one linear evolutionary development. Rosenfeld also expressed concern about the state of the history of science that the social aspects were so little considered. He concluded his introduction with the following:

> One may think that it is a bad sign that a branch of science should be still uncertain about its method. On the other hand, it is certainly a good sign that the significance of the social relations of science, not so long ago utterly ignored by the majority of historians and scientists, is now fully recognized, and that their historical development is explored with an energy promising fruitful results. (Rosenfeld 1953, 11)[7]

Centaurus volume 3 (1953), particularly Rosenfeld's introductory remarks, was forward-looking in terms of the direction of the history of science. But generally speaking, the narrative influenced by trends in the

[7] It is interesting to note that Rosenfeld explained the difficulty in selecting authors and articles to represent more globally the science–society relationship. Thus, "It proved impossible . . . to find competent scholars to supply all the desired contributions, especially in the wide and little-explored field of oriental science" (1953, 9). He apologized for the omission of Arabic and Indian civilizations in the volume's representation, saying, "Much against our wish, our selection, therefore, perpetuates the prevailing bias towards occidental culture which to no negligible degree distorts the picture of the growth of human culture and obscures the striking unity of this growth" (9).

heavily demarcationist philosophy of science of the early to mid twentieth century would prevail until the so-called postpositivist turn of the later twentieth century. As a matter of the history of the historiography of science during the twentieth (and already nineteenth) century, then, it is important to see that a powerful philosophical attitude about the epistemological and methodological boundaries of science accommodated discrepant accounts of early science, such as we see in Neugebauer's *Exact Sciences in Antiquity* and Pedersen's *Early Physics and Astronomy*.

While the demarcation problem was retooled by modern philosophers of science, it invariably had an impact on historians of science, whether historians cared to weigh in on the question expressly or not; and most did not. The demarcationists' essentially nonhistorical, even ahistorical, effort focused on modern, not historical, sciences. Their criteria, therefore, act as a distorting factor when adopted for the history of ancient science. Although Neugebauer expressed the difference between rational astronomy and irrational astrology, he also said in his "Study of Wretched Subjects" that "the foundation of our studies" is "the recovery and study of the texts as they are, regardless of our own tastes and prejudices" (1951, 111).

Against what was still in the early 1960s an effective narrative about the nature of science and its development, Kuhn suggested that the uniqueness and singularity of science be viewed in the face of a broad range of historical evidence, that is, broader than the mathematical sciences, which were more readily assimilated to a modern image. History, therefore, as Kuhn stated in the opening to *The Structure of Scientific Revolutions*, should be transformative for the image of science. From this perspective, it became possible, and wholly acceptable, to approach ancient and medieval magic, astrology, and alchemy as legitimate scientific programs within their own historical contexts, regardless of their status as irrational in ours. The confrontation with a broader range of historical sources as a result of a newly historicized image of science, in turn, introduced questions about the unity of science, its image as a mirror of the reality of nature, or its purpose as the epistemological method to discovering the truth about nature.

In terms of the definition, purpose, and image of science, the histories of Neugebauer and Pedersen differed in certain significant ways in reference to the ancient Middle Eastern contribution. Neugebauer rooted science in Babylonian astronomy, seeing one unbroken continuity from Babylonian antiquity to early modern Europe in the application of mathematics and quantitative methods to solving problems of the phenomena. In line with modern scientific values, mathematical astronomy represented a historical continuity in its concern to produce astronomical theory (lunar and planetary), that is, general techniques of computation suitable for application to a broad range of particulars. Pedersen, like Schrödinger, took science back only to the Greeks and maintained the idea that Babylonian astronomy was merely practical or too religious.

The second major difference in the historiographical approach between Neugebauer and Pedersen shows in their respective attitudes toward philosophy, both ancient and modern. Neugebauer did not consider philosophy to be a legitimate contributor to the progress of science but rather a hindrance as he evaluated its role in the Greek milieu. Pedersen considered philosophy as a motivating and creative force, contributing to a rigorous rationality.

These differences aside, Neugebauer and Pedersen shared a core view of science as the investigation of the physical phenomena of nature, and as such shared a basic commitment to the continuity of science from its origins (whether in Babylonia or in Greece) to the modern era. As far as the demarcation between science and nonscience was concerned, Neugebauer took greater pains to defend the relevance of astrology to the history of science as the great conveyor of astronomical predictive methodology, although his indirect and narrow defense of astrology as science in his *Exact Sciences*, consistent with his position in "The Study of Wretched Subjects," was mounted within the bounds of the idea of the unity of science and his characterization of astrology as a mechanistic system. Pedersen adopted a harder line with reference to the rationality of astronomy and the irrationality of astrology. Issues of demarcation were important to both, but for different reasons.

Finally, it must be said that at the time of the writing of *The Exact Sciences in Antiquity* and *Early Physics and Astronomy*, there had been very little attention paid to the details of the ancient exact sciences, particularly of astronomy. These works brought material that was all too poorly known to the attention of a greater history of science public. Their differences point to the tensions that lingered over the long-held conviction that science was invented by the Greeks. This point, so central to the historiography of early science of that time, was able to be moved off center through the works of Neugebauer, not only his *Exact Sciences in Antiquity* but the three-volume *Astronomical Cuneiform Texts* and the three-volume *A History of Ancient Mathematical Astronomy* as well. The foundational work of Neugebauer combined with the historicizing trends of the postpositivist period up to and including today have materially changed the historiography of science. With this change has come our ability to see the relevance of Middle Eastern antiquity, specifically the cuneiform world, to the overall picture of the history of science, not simply as a rudimentary, undeveloped, or less perfectly scientific prefiguration of Greek science but as a fully independent tradition with its own line of kinship with modern exact science.

DAVID PINGREE AT BROWN UNIVERSITY

Twenty years after Neugebauer's "Wretched Subjects," in the year 1971, David Pingree joined the History of Mathematics Department at Brown University. The department was established by Neugebauer in 1947 and dissolved in 2005. Another founding member, coming in as Neugebauer's assistant in 1941, was the Assyriologist Abraham J. Sachs. The History of Mathematics Department would invite for various terms of residency numerous scholars who were among the very most important historians of ancient and medieval scientific materials from the Middle East, the Mediterranean, and Europe (e.g., Gerald Toomer, Bernard Goldstein, Noel M. Swerdlow, and E. S. Kennedy).[8]

The Brown group, including its many visitors and scholars in residence, did not have a conscious investment in a unified school of thought or historiography. Neugebauer and Pingree underpinned their respective historiographical positions differently, as will be brought out in this section focusing on Pingree. The heterogeneity of the members of Brown University's small department of specialists in the astral sciences of the past – astronomy and astrology in the ancient Near East (including Egypt), India, and the Greco-Roman worlds – only lends that much more of a notable profile to that department as it existed up to the death in 2005 of David Pingree, its last surviving member in residence.

Pingree, more than anyone else, would investigate and make available the sources for the study of how astronomy and astrology together functioned in a variety of contexts across the Hellenistic *oikoumene*, up to and including India, and in the late fourth and fifth centuries of our era in Sasanian Iran, and from there too into the creative period of Arabic astral sciences. In one of Pingree's editions of Sanskrit astrological works, the *Yavanajātaka*, or "Horoscopy of the Greeks," he noted that this author's

[8] A detailed summary of the History of Mathematics Department at Brown, including its distinguished faculty, visitors, and students, may be found at www.wilbourhall.org/History%20of%20the%20History%20of%20Math%20Dept.htm.

work was "concerned with horoscopy, or *horaskandha*," which was "one of the three branches (*skandhas*) of *jyotihśāstra* (astronomy and astrology)" (1978, 1.5). He assessed the *Yavanajātaka* to be the earliest surviving Sanskrit text on horoscopic astrology,[9] arguing that it "constitutes the basis of all later Indian developments in horoscopy before the injection of Islamic theories through translations from Persian into Sanskrit of *tājika* works in the thirteenth and the following centuries" (Pingree 1978, 1.5), and he noted that "the second skandha (branch) of *jyotihśāstra*, called *samhitā*, goes back to a Mesopotamian tradition similar to that of the great omen series, *Enūma Anu Enlil*" (Pingree 1978, 1.5).

In the mid 1970s, Pingree continued the work of editing Hellenistic astrological treatises. In 1974, he reedited the early fifth-century CE Hephaistio of Thebes' *Apotelesmatica*, which had been published in its first edition by August Engelbrecht in 1887; in 1976, the first-century CE Dorotheus of Sidon's *Carmen Astrologicum*; and in 1986, a reedition of Wilhelm Kroll's 1907 first edition of the *Anthologium* of Vettius Valens from the second century CE. Thanks to these and many other text editions, we now have the materials to pursue further study of the ways astrology and astronomy of widely varying origins circulated in widely disparate societies.

Finally, it was in bringing Babylonian scientific texts from the divinatory (and astral magical) tradition within the framework of Hellenistic Greek, Greco-Roman, and also East Asian tradition that Pingree became a central figure in the historiography of the cuneiform astral sciences. His collaborations with Erica Reiner (1975, 1981, 1998, 2005) on the editions of the planetary and fixed-star omens of the series *Enūma Anu Enlil*, as well as his historical studies involving cuneiform texts and their transmission to the east (1992b, 1997, 1998), provided the basis for his cultural approach to the history of astronomy and astrology.

The impact of Pingree's approach to the celestial sciences in antiquity cannot be overestimated. He brought to the fore the multiple forms and contexts science took in premodern times. He said:

> One of the most significant things one learns from the study of the exact sciences as practiced in a number of ancient and medieval societies is that, while science has

[9] Further research on this text by Bill Mak (2013a, 2013b, 2014) and others (cited in Mak 2013a and in Brown et al. 2018, 215 and 675) makes the dating of this text open to question and doubt, and there has been considerable criticism of Pingree's edition and interpretations. None, however, can take away from the pioneering position of his work and the fact that his work on Indian science has been a springboard for many younger scholars now established in the field, such as Clemency Montelle, Kim Plofker, and Toke Knudsen.

always traveled from one culture to another, each culture before the modern period approached the sciences it received in its own unique way and transformed them into forms compatible with its own modes of thought. Science is a product of culture; it is not a single, unified entity. Therefore, a historian of pre-modern scientific texts – whether they be written in Akkadian, Arabic, Chinese, Egyptian, Greek, Hebrew, Latin, Persian, Sanskrit, or any other linguistic bearer of a distinct culture – must avoid the temptation to conceive of these sciences as more or less clumsy attempts to express modern scientific ideas. They must be understood and appreciated as what their practitioners believed them to be. (Pingree 2003, 45)

In his 1992 article "Hellenophilia versus the History of Science," Pingree (1992a) took his perhaps most strident historicist stance with regard to the sciences of other cultures and delivered his clearest indictment of Western historians who failed to appreciate the scientific nature of non-Western bodies of knowledge. In what is in effect a response, however indirect, to Schrödinger's claim that science was the "Greek way of thinking," Pingree coined the term Hellenophile (as opposed to a Philhellene), to represent

someone who suffers from a form of madness that blinds him or her to historical truth and creates in the imagination the idea that one of several false propositions is true. The first of these is that the Greeks invented science; the second is that they discovered a way to truth, the scientific method, that we are now successfully following; the third is that the only real sciences are those that began in Greece; and the fourth (and last?) is that the true definition of science is just that which scientists happen to be doing now, following a method or methods adumbrated by the Greeks, but never fully understood or utilized by them. (Pingree 1992a, 555)

Pingree did not set himself up as a relativist, nor did he have any stakes in a philosophy of science. Very much in line with Neugebauer's view that texts were "the foundation of our studies" (Neugebauer 1951, 111), Pingree's emphasis was also on the sources, and the Sanskrit, Arabic, and Akkadian texts that occupied him shaped his perspective on historical sciences and their cultural integrity.

Beyond this cursory glance at the contributions of Neugebauer and Pingree, it is possible, despite their mutual support and intellectual sympathies and their agreement on placing emphasis on the recovery and edition of primary sources, to infer quite different approaches to the question of science as a historical phenomenon. As discussed in the previous section, Neugebauer approached ancient astronomy from the perspective that science and scientific knowledge comprised a historical and epistemological unity, while Pingree's approach favored disunity, comprising a plurality of cultural practices and epistemologies.

The historiographical orthodoxy of the unity of science, when Schrödinger was writing *Nature and the Greeks* and Neugebauer was preparing ACT, is less convincing in light of the textual materials for astrology and magic edited by Pingree. Pingree's historiographical legacy generated a wealth of new work in the premodern non-Western sciences. It would be far beyond the present scope to list a full bibliography, but for the scientific textual cultures of astronomy, astrology, divination, magic, medicine, physiognomy, and technology of the ancient Middle East, the following are indicative of this development seen in works by Egyptologists (von Lieven 1999, 2000, 2006; Quack 2018a and 2018b, 2022), Sanskritists (Yano 1994–2019; Plofker 2008; Montelle 2011; Montelle and Plofker 2018), Iranists (Panaino 2021), Semitists (Ben-Dov 2008, 2014, 2018; Ben-Dov and Sanders 2014; Drawnel 2011; Bohak and Geller 2013), and Assyriologists (Brack-Bernsen 1969–2021; Geller 1999–2022; Steele 2002a–2022; Heeßel 2005–2021; Ossendrijver 2011a–2021b; de Zorzi 2014, 2017; Johnson and Stavru 2019; Borrelli and Escobar 2022; Escobar 2017, 2018, 2019a, 2019b, 2022, 2023). Contemporary work in the history of premodern sciences of the Chinese (Lloyd and Sivin 2003; Hsia 2009, 2017; Chemla 2012; Mak 2013a, 2013b, 2014, 2019) and Islamicate worlds (Brentjes 2010, 2018; Brentjes and Renn 2016) furthers expands the cultural landscape and diversity for a new historiography of the premodern non-Western sciences (see also Bretelle-Establet 2010).

This recent research in traditions of astrology, alchemy, magic, medicine, divination, and physiognomy across the Middle East and Asia from antiquity to the beginning of early modernity creates a place in the historiography of science for a cultural and contextual approach that can be seen as the legacy of David Pingree. A historiography of science that further takes into account culturally diverse objects of scientific inquiry, defined in terms of what a particular community perceived as its phenomenal world, can begin to entertain questions about the cultural perception of worlds and in so doing enter into the realm of the anthropology of early science.

PART II

WORLDMAKING AND THE ANTHROPOLOGY OF SCIENCE

We have gradually come to accept that the epistemic values of antiquity were utterly different from our own. The ideas of nature, or no nature, or the gods, or nature as God's creation form distinct parts of conceptual matrices within which to frame our inquiries into the ways ancient cultures approached the phenomena comprising ancient medicine, astronomy, astrology, and the many scholarly forms of divination from signs. Once we have restored the cuneiform sciences to their own broad-based intellectual historical contexts with their distinct epistemic goals, the question then arises as to the character of the objects of cuneiform scientific inquiry. As initial steps in answering this fundamentally ontological question, Part II (Chapters 4–7) takes up the notional complex of worldviews, world-versions, and worlds.

The discussion in Part II is framed by already well-established discourses in other disciplines that have long reflected on such questions of classification, representation, and ontology, principally from philosophy, philosophy of science, and cultural anthropology. These ideas will underpin the examination of the Babylonian classification, modeling, and description of phenomena and features of their world. Beyond Babylonia, this framework has clear relevance for an anthropology of science centered on the intellectual and practical engagement of premodern and non-Western cultures with their worlds.

4

The Idea of Worlds

> Languages that have evolved in distant times or places may differ extensively in their resources for dealing with one or another range of phenomena. ... But examples like these, impressive as they occasionally are, are not so extreme but that the changes and the contrasts can be explained and described using the equipment of a single language.
>
> – Donald Davidson, 1973

What does it mean to say that the world referred to in historical texts is not "the same" as some other world, say that of our own, or the one that science comes to know over time? At first blush, the idea of worlds seems wrong, counterintuitive, even illogical, because the word "world" is used to refer to the one encompassing totality of everything that exists. If it were only a matter of knowing *the* world, assuming a common and universal point of reference, a touchstone by which to judge whether what has been known was on the right track (Babylonian astronomy) or simply in error (Babylonian astrology), then talk of many worldviews would suffice. What then is the reason to consider the idea of worlds rather than worldviews, and what work does the idea do for the historiography of ancient science?

The chapters of Part I touched upon the problems with the retrospective approach to the history of science shared by presentism, Whig histories of science, the pessimistic (meta)induction, and the demarcation between science and nonscience for periods before modernity. None of these approaches focused on science as a cultural phenomenon, and all maintained a largely epistemological focus. A historiographical counterpoint then redirected its questions to how science emerged and took certain forms as a function of the cultural interests of its practitioners. This was the general attitude of postpositivist social historians of science and contextualists following Ludwik Fleck, Alexandre Koyré, Thomas Kuhn, and Quentin Skinner. The cultural approach is distinct in that its premise is the claim that at the core of science, as well as every other product of culture regardless of its geocultural or temporal location, is an image and understanding of a world inhabited by those in it as a matter of reality. But does evidence of those many historical realities justify the idea of a plurality of worlds? Are we speaking only rhetorically?

Babylonian intellectual culture mounted the longest recorded astronomical observational program to have ever existed. Nearly 800 years of regular nightly watching of heavenly phenomena at Babylon from the eighth century BCE to the middle of the first century BCE (Sachs–Hunger–Steele ADART) was recorded and compiled into a systematic archive and arranged by date. Modern scholarship has termed this archive the Astronomical Diaries. The phenomena on which the Diaries focused were principally the moon's position with respect to the stars, the planetary positions and dates of their synodic phenomena, and many other astronomical and some nonastronomical data (market prices, river levels, odd events in the city streets, military activity worth reporting). On the basis of this archive of mostly astronomical observations, we feel a connection, distant though it certainly is, to the Babylonian scribes for their development of a systematic empirical basis for astronomy.

However, do we not assume too much if we say that the frame of reference for Babylonian observation, measurement, and calculation of the regularly appearing celestial phenomena was the same as that of our scientific astronomy? Perhaps in the sense of the limited framework of positional astronomy, such as we find in the astronomical tables of the *Nautical Almanac*, we can assume "the same" framework. Evaluating the quality of the Babylonian observations may rest upon the assumption that they observed the same phenomena, but how they understood the phenomena, for example, planetary synodic phenomena, apart from their positions with respect to the fixed stars or the zodiacal signs is a different question. What the stars were thought to be and where they were fitted within a systematic whole are ontological questions left unanswered by the observational archive of the Diaries but can be explored in other genres of cuneiform texts. If as historians of science we ask the question of whether the world to which our historical sources refer is the same as that of modern (or some other) science (as Lloyd 2006, 76 does), we should consider that question in its fullest sense as consisting of both the epistemological (knowledge of what there is and how it is acquired) and the ontological (what there is), the ontological dimension being perhaps the more fundamental.

A brief discussion of the terms and stakes of the ontological part of this question was given in the Introduction. Here, I return to those issues to discuss some key crosscurrents through the notional territory of world-views, world-versions, and worlds from philosophy, science and technology studies (STS), and anthropology. These will serve to situate the question of the idea of worlds (pluriworldism) and its implied conceptual relativism before considering the cuneiform scribes' descriptions of the world and its phenomena (Chapters 5–7).

THE ONTOLOGICAL TURN

Ontology has had a long career in philosophy, going back to Parmenides, to Aristotle and the Scholastics who dealt with his ideas, and then to a renewal in the modern philosophy of science. Without rehearsing the great debate between nominalism and realism here, which is well beyond the purview of this book, the old problem of the "existence" of things, whether *concreta* (particulars) or *abstracta* (universals), came to be of central importance to twentieth-century logical empiricists who defined science as knowledge expressed through statements about physical, not metaphysical, things. Embedded in the new ontology was a critique of ancient and medieval metaphysics, which had considered questions on the reality of transcendent entities.

Twentieth-century philosophy of science emerged from this critique and took a new turn with Nelson Goodman and Willard von Orman Quine (Goodman and Quine 1947), who established an even more rigorous nominalism based on a renunciation of abstract entities (Gosselin 1990, 10–21), although ultimately they differed on what entities had ontological privilege; for Quine, it was concrete physical things, for Goodman, individuals (Goodman 1972, 155–72), without specifying what those individuals are taken to be. What is most significant for the present context is that the new postwar period ontology of the Goodman–Quine variety, was, in Gosselin's words,

> no longer absolute, but *relative to a definite language*. Thus there are as many ontologies as there are languages. It *defines the universe the language under consideration can describe*, provided we can interpret that theory in another broader theory, which can be reinterpreted in a broader theory, until we reach an ultimate background theory with "its own primitively adopted and ultimately

inscrutable ontology." (Gosselin 1990, 16, quoting Quine 1969, 51; emphasis in Gosselin)

Quine's ontological relativity (1968, 1969) had to do with "the inscrutability of reference," and so ontology was viewed as relative to a certain language or use of language (such as in science or something else).

Thus "A question of the form 'What is an F?' can be answered only by recourse to a further term: 'An F is a G' [F and G are predicates]. The answer makes only relative sense: sense relative to an uncritical acceptance of 'G'" (Quine 1968, 209). In this way, Quine and Goodman's nominalism and empiricism had profound implications for constructivist (or, as some call it, constructionalist) ontology and therefore ontological relativism. Gosselin noted: "If ontology is the answer to the question 'what is there?,' namely 'everything,' the universe that can be described in the constructivistic language, in principle can contain such diverse entities as 'electrons,' 'plants' and 'good and bad'" (Gosselin 1990, 18). One significant difference between Quine and Goodman on ontological relativism is that Quine's relativity of reference did not entail multiple worlds as did Goodman's.[1] Quine's "ultimately inscrutable ontology" (Quine 1968, 202) also seems incompatible with Goodman's thoroughgoing constructivism in which even the ultimate realities must be relative to some system of reference, scrutable in terms of some system.

While Quine's ontological relativity (1953, 1968, 1969) loomed large in twentieth-century analytic philosophy, anthropologists had different stakes in understanding the varying ontological commitments of other cultures. Pioneers in this from the field of anthropology are Bruno Latour, who spoke of variable ontologies (1993, 85–88; 2009); Eduardo Viveiros de Castro, who spoke of perspectivism (1998, 2004a, [2004b] 2019, 2012, 2013); and Philippe Descola, who spoke of "worldings"[2] or ontological regimes (2010, 2013, 2014). As a counter to the ontological pluralists, João de Pina-Cabral (2014a and 2014b) argued for what he called a "minimalist realism," in which human beings engage in diverse

[1] Geoffrey Hellman's introduction to the 3rd edition of Goodman, *Structure of Appearance* (1977), xxxviii–xxxix discusses the relation of Goodman's ideas to those of Quine.

[2] Descola noted that his use of the term "worlding" is not that of Gayatri Spivak and other postmodernists (2010, 339 footnote 1). "Worlding" in Spivak's usage, which goes back to Heidegger, is, to quote Deepika Marya, "a reference to the construction of the world through a geopolitical economy that enables us to configure and determine the shape of that world" (Marya 2017, 165). Descola clarified his meaning of worlding as against that of "the one given to that word by postmodern and postcolonial authors, that is, as a social construction of reality by hegemonic Westerners" (Descola 2014, 272).

ways with only one world and a "single-world" ontology (Pina-Cabral 2014a, 51–52).

In a two-part article series, Pina-Cabral (2014a and b) raised the question as to whether "world" was still a viable category for anthropologists. He noted that "the perplexities" about the usage of the word "world" "have haunted the social sciences for a very long time" (2014a, 50). Pina-Cabral described the twentieth-century discourse on worlds as ranging from realism to idealism.[3] He characterized perspectivism, of the kind conceived by Viveiros de Castro, as metaphysical reenchantment taking over in the 2000s (Pina-Cabral 2014b, 150–51). Pina-Cabral's concern was less philosophical and more ethnographical. He was concerned to define the concept and use of the word "world" in contemporary anthropology in order "to propose an outlook on world that allows both for the universalist hopes of the anthropological endeavor and for the particularist demands of the ethnographic practice" (2014a, 50–51).

Descola's worldings (2010, 336–39; 2014, 272–73) offered a variant on ontological plurality. His ideas differ from those of Goodman's worldmaking and even stem from opposing claims. Worlding, by Descola's account, is the result of the stabilizing effect of "ontological predication," that is, what people say things are (x is F; Descola 2014), which resonates with Quine's ontological relativism in its attachment of ontology to language. His idea is also like Quine's in that it does not lead from ontological relativism to pluriworldism.

Descola made clear that he does not subscribe to pluriworldism, saying, "My position excludes both the hypothesis of multiple worlds and that of multiple world-views" (Descola 2010, 339). No multiple worlds, he said, "because it is highly probable that the potential qualities and relations afforded to human cognition and enactment are the same everywhere until some have been detected and actualized, others ignored" (Descola 2014, 277). No worldviews, because

> once the worlding process has been achieved, once some of these qualities and relations have been detected and systematized, the result is not a worldview, that is, one version among others of the same transcendental reality; the result is a world in its own right, a system of incompletely actualized properties, saturated with meaning and replete with agency, but partially overlapping with other similar

[3] Another opponent of the ontological turn, David Graeber accused the anthropological ontologists of simply turning ontology into idealism, not being sure "whether they are trying to dissolve Materialism into Idealism, Idealism into Materialism, or both Materialism and Idealism into something else" (2015, 21, note 24).

configurations that have been differently actualized and instituted by different actants. (Descola 2014, 277–78)

In the first passage, variant worldings refer to one and the same world. In the second passage, the claim that the result of worlding is not a worldview "but a world in its own right" may seem to come perilously close to worldmaking and pluriworldism, but it does not, because Descola defines the ontological stuff of systems in a different way from that of Goodman's worldmaking.

According to Descola, "the material and immaterial objects of our environment do not stand in the heavens of eternal ideas ready to be captured by our faculties, nor are they mere social constructs giving shape and meaning to a raw material; they are just clusters of qualities some of which we detect, some of which we ignore" (Descola 2014, 273). This suggests an inchoateness to the stuff of ontologies, which is itself not brought into existence wholly by human cognitive activity but takes shape and form depending on the preferences, habits, or "filters" (Descola 2014, 273) of different cultural systems. The problem of what is made versus what is found is, however, left unresolved.

Descola's worlding has its own character in the account of multiple ontological organizations of universally available ontological potentialities (Descola 2010, 2013, 2014). Worldings are multiple in the sense that they are based in four heuristic ontological schemes, decidedly not the equivalent of multiple actual worlds, but rather "fragmentary actualizations" of universally available ontological things/properties. There is an element of Descola's formulation that resonates with Goodman's worldmaking, which does not posit worlds made up out of "whole cloth" or from conjuring features out of a featureless "uninterpreted reality." In order to circumvent the charge of unrestrained fabrication, Goodman's worldmaking comes of an interdependent relationship between finding and making, and the remaking of inherited ways of worldmaking. I shall return to this issue in the section "Pluriworldism and History." Suffice it to note here that, however much Descola's worlding may share something with worldmaking, he is neither irrealist nor pluriworldist. The chief advocates of pluriworldism, both Goodman and Kuhn, struggled to account for whence the stuff of the making of worlds came in the cognitive process of making, and the chief critic, or one of them, Israel Scheffler, insisted that the claim to such making was rhetorical, not literal.

As indicated in the Introduction, signs of the ontological turn emerged as well in STS, where in 2012/13 the journal *Social Studies of Science* brought out a special issue (volume 43) devoted to the topic with the title "A Turn to Ontology in Science and Technology Studies?" In a topic search on the Web of Science from 1989 to 2008, Bas van Heur, Leydesdorff, and Wyatt (2012, 355) saw an enormous increase in the occurrence of the term "ontology," but they questioned whether that evidence signaled a true turn in STS in the form of a shift in theoretical or metaphysical orientation. What the increase in ontological vocabulary pointed to in their analysis was rather a broader field of interest in ontology for STS, both in picking up on old debates about constructivism versus realism and adding to these existing debates further discussion on classification, instruments, and technologies involved in scientific practice.

In the same volume, Steve Woolgar and Javier Lezaun (2013) assessed the emergence of the ontological discourse within STS against the background of philosophy and anthropology. They saw in the work concerned with ontology a change of course away from interpreting "multiple worlds" as the multiple ways of representing one world, in other words a separation of multiple worlds discourse from that of worldviews. Another paper in the volume (Lynch 2013, 444) offered to add to the existing STS ontological lexicon the term "ontography."[4] Ontography was meant to cover, essentially, ways of worldmaking in actual practice concerning certain bodies of evidence. Michael Lynch said:

> Currently in STS, moves to pluralize "epistemologies" in social, historical, and cultural studies of science and technology are more ubiquitous than any single worked-out epistemological program. For quite some time, "epistemologies" have been identified with distinct standpoints in history and society (Haraway, 1988). Understood as an STS research program, epistemology – and, by extension, ontology, ethics, or aesthetics – does not make up a distinct field of metaphysics, but instead, it reverts to diverse social, historical, political, and cultural conditions under which "knowledges" are established, objectified, moralized, communicated, or dismantled. (2013, 451)

The sociology of scientific knowledge, the springboard for STS, according to Lynch was still interested in credibility contests, that is, in what is defined as knowledge in certain historical contexts or ceases to count as knowledge in others (2013, 451). Knowledge, as an empirical subject

[4] Lynch explained that when he first introduced the term in 2008 at the Oxford Ontologies Workshop, he had not realized that others had already used the neologism for exactly the metaphysical position he had meant to contrast with his own usage (2013, 460). Martin Holbraad also used the term in the 2010 forum on "Worlds Otherwise," published in Alberti et al. 2011, 908.

matter, just like science in the Kuhnian sense, is only knowledge as a function of the communities that develop and preserve those traditions or theories; thus we have evidence of a plurality and disunity of knowledge and a plurality and disunity of science.

Whether the plurality and disunity of ontologies is an extension of this plurality about knowledge and science, however, is another matter; it does not automatically follow, and so the question of one world or many does not rest with the evidence of epistemic plurality. All in all, *Social Studies of Science* volume 43 did not reflect a unified turn to a particular approach to a definition of ontology for STS that would unify a position on whether evidence from Science Studies constitutes diverse representations of one ontological reality (Lynch 2013, 445 summarizes). In sum, no consensus obtained either for a plurality of ontologies reconcilable within a single world, or a plurality of ontologies as making many worlds.

More recently, anthropologists and historians of science, convening in Cambridge around the subject "Science in the Forest, Science in the Past," have advanced the project of the ontological turn in their respective disciplines.[5] Nick Jardine discussed what he called "distant sciences and the ontological turn," saying:

> Proponents of the ontological turn are almost unanimous in breaking with the nature/culture division. In doing so they break with both strong realism, belief in the existence of a single coherent and exclusive truth about the world, and strong relativism, belief in the existence of many equally valid but radically different perspectives on the world (Paleček and Risjord 2012). Instead they are pluralists, prepared to recognize many worlds or many "dimensions" of reality. (Jardine 2019, 173)

Indeed, Martin Paleček and Mark Risjord, writing in a special Roundtable on the Philosophy of Social Science in the journal *Philosophy of the Social Sciences* (volume 43), discussed a new form of relativism that suited those interested in an ontology that no longer functioned as transcendental truth or substance, inertly waiting "out there" for human beings to interpret it in diverse ways in the form of diverse worldviews.

Those interested in the new ontology were anthropologists seeking to explore the ways cultures conceive of the nature of things, their substance, their meaning, and their reality. "Proponents of the ontological turn therefore see the representationalist standpoint of modern philosophy," Paleček and Risjord said, as underpinned by the "problematic ways in

[5] The first meeting, in 2017, was published in *HAU: Journal of Ethnographic Theory* (2019) and the second in *Interdisciplinary Science Reviews* (2019) in McCarty et al. 2022.

which culture was conceptualized in the 20th century. By rejecting representationalism, they hope to inaugurate a deeply different approach to cultural anthropology" (Paleček and Risjord 2012, 7). The representationalists, as well as the interpretivists (e.g., Clifford Geertz), were interested in what symbols and their structures denoted of the things that furnish the world, whereas the ontologists sought to collapse the symbol and its object and make concepts and things one (Henare, Holbraad, and Wastell 2007, 13). Things have meaning through context and usage rather than by symbolism and representation.[6]

Paleček and Risjord saw this brand of ontological interest breathe new life into the old conceptual relativism debate that had arisen following Donald Davidson's 1973 paper "On the Very Idea of a Conceptual Scheme" (later republished in his 1984 monograph *Inquiries into Truth and Interpretation*). I return to this aspect of the ontological implications of Davidson's conceptual schemes in relation to Goodman's worldmaking in the section "Conceptual Relativism and Translation."

Geoffrey Lloyd's introduction to the second collection of papers from "Science in the Forest, Science in the Past" (McCarty, Lloyd, and Vilaça [2021] 2022) makes the link I emphasize here, namely, between historians and anthropologists: "We are all faced with radical otherness, whether we are ancient historians or modern ones or anthropologists. That otherness takes different forms, posing different challenges to our understanding and it does seem important to recognise those differences. Alternative customs are one thing, values another, ontologies, science, mathematics yet others" (Lloyd 2021b, 223). Beneath the wide variety of interpretation by historians and anthropologists of non-Western systems of knowledge and their ontological underpinnings is invariably the question of how different cultural knowledge systems or "worldviews" stand in relation

[6] This might be an interesting solution to the problem of metaphor in Akkadian terms for celestial phenomena (Rochberg 1996, 2010c). Instead of seeing the moon (the metaphoric subject) and the moon god (the metaphoric vehicle) as two distinct things, an object and a symbol, where the metaphor refers to the symbol (the god) as "representing" the moon, as in the moon god's dark face being a *symbol* of an eclipse, we might see that the moon is the celestial disk, or other shape, in the sky in some contexts, a moon god with feelings in other contexts. One may well ask, however, whether they are never distinct or independent from one another *à la* Henare, Holbraad, and Wastell (2007, 14)? The use of metaphor is a common mode of describing things, one which requires separation between the subject and the vehicle to effect the force of figures of speech as well as figures of thought. This is not the same as collapsing "appearance and reality" (Henare, Holbraad, and Wastell 2007, 14). In accordance with Goodman's constructivism, "reality" is as conceptualized by frames of reference as are the things that furnish worlds. Worldmaking neither obviates metaphor nor collapses objects into concepts.

to the world, or whether, indeed, different systems (world-versions) make different worlds. This was the question at the heart of Goodman's *Ways of Worldmaking* (1978), to which I now turn.

WORLDVIEWS, WORLD-VERSIONS, AND WORLDS

More directly relevant for the historical-anthropological project than Quine's ontological relativity, which did not support pluriworldism, was the philosophy of his contemporary, Nelson Goodman. Goodman's *Ways of Worldmaking* (1978) was itself the culmination of his previous books: *The Structure of Appearance* ([1951] 1977), *Fact, Fiction, and Forecast* ([1954] 1983), *Languages of Art* (1968), and *Problems and Projects* (1972). Following considerable criticism (e.g., Scheffler 1980; 2009, 66–72),[7] Goodman's later works "Notes on the Well-Made World" (1983), incorporated within *Of Mind and Other Matters* (1984), addressed and clarified many points.

Goodman's thoroughgoing constructivism had already given the philosophical community pause before *Ways of Worldmaking*, as attested in the editorial preface to *Structure of Appearance* ([1951] 1977). The series editors Robert S. Cohen and Marx Wartofsky said there about Nelson Goodman that "his bite is worse than his bark" (vii). They elaborated: "Behind what appears as a cool and methodical analysis of the conditions of the construction of systems, there lurks a radical and disturbing thesis: that the world is, in itself, no more one way than another, nor are we. It depends on the ways in which we take it, *and on what we do*" (vii; emphasis in the original). This notion of the world, as not being "in itself" any particular intrinsic "way," is the premise for Goodman's irrealism (see Introduction for further discussion of Goodman's irrealism).[8]

[7] Scheffler also accused Kuhn of "extravagant idealism" in *Science and Subjectivity* (1967, 19) in the chapter "Objectivity under Attack." Note the softening of his position on pluralism, though not realism, in his articulation of "plurealism" (Scheffler 2009, 62–66).

[8] Goodman was not the first to approach the idea of the world in terms of a radical ideational and linguistic construction. Although on diametrically opposing points on the spectrum from realism to irrealism, a prominent member of the Lvov–Warsaw School in the 1930s, Kazimierz Adjukiewicz, developed a notion of "world-picture" or "world-perspective." In a 1934 paper, Adjukiewicz said, "Of all the judgments which we accept and which accordingly constitute our entire world-picture, none is unambiguously determined by experiential data; every one of them depends on the conceptual apparatus we choose to use in representing experiential data. We can choose, however, one or another conceptual apparatus which will affect our whole world-picture" (Adjukiewicz [1934] 1978, 68). This relationship of ideas has been brought out by Peter Olen in connection not with Goodman but with Wilfrid Sellars (whose pragmatic realism employed the terms world-perspective

The irrealism of *Ways of Worldmaking* was present already in *Structure of Appearance*. While philosophical factions duked it out over physicalism, realism, phenomenalism, and antirealism, Goodman held them all to be conventional and none more apt than another. He compared the reality we see in our world to realism in art, finding both "largely a matter of habit" (1978, 20). Of all the *right* ways the world was constructed, it was "no more one way than another." Indeed, it was either the one way we take it, all the ways we take it or have taken it, or no way, in a coalescence of pluralism, nihilism, and monism (Goodman 1983), all a matter of convention.

In Goodman's view, convention, generally understood as "fabricated form imposed on uninterpreted content," was a meaningless construct for the reason that he found the notion of uninterpreted content to be itself unintelligible and, as a result, the disconnection between convention (as arbitrary) and content (as factual) to be ill conceived (Goodman [1989] 2010, 80–81).[9] Facts he saw as "creatures of their descriptions," impossible to disjoin, much as world-version and world were interdependent. "All convention," he said, "depends upon fact, yet all fact is convention" (Goodman [1989] 2010, 82). It was in terms of this sense of the conventional status of all the ways of worldmaking that Goodman defined irrealism, not as the denial of reality. Irrealism did not deny the reality of anything but made descriptions of reality (world-versions) a matter of relation to a framework. In turn, the framework, in his view, stood in complete interdependence with the world.

The evanescent ontology of Goodman's philosophy, wherein worlds "melt" into versions, and experience and theory are part of the worldmaking process (Goodman 1984, 29 and throughout), resonates both with the idea of the construction of facts and the relation between facts and artifacts dealt with by Latour and Woolgar ([1979] 1986, 176–83). Hacking (1988, 282) likened Latour and Woolgar's approach, not to the discourse on realism versus antirealism but to Goodman's irrealism, according to which versions make worlds. Not to be conflated, versions and worlds have an evanescent unfixed relationship. Worlds contribute to

and world-story), and Sellars's influences, which included Quine. For Olen's discussion see Olen 2019, 55–74. Each of these formulations, of Adjukiewicz, Sellars, Quine, and Goodman are distinct in what they are committed to. My mention of them together here is only to point out that the question of the relation of language to world has a long, complex, and varied history in twentieth-century philosophy and, with respect to notions of "world," certainly deserves a full review in another context.

[9] Cf. Davidson's critique of scheme–content dualism in Davidson 1973, 11; 2005, 48.

and ultimately answer to versions produced by whatever worldmaking process is operative, be it science or art. Latour and Woolgar said, "We do not wish to say that facts do not exist nor that there is no such thing as reality. . . . Our point is that 'out-there-ness' is the *consequence* of scientific work rather than its *cause*" ([1979] 1986, 180–82). In accordance with irrealism, the found (out-there) and the made (by means of science) are conjoined in a dynamic interrelationship (see note 17).

Hacking (1988) was right to see the resonance between Latour and Woolgar's position and Goodman's irrealism and worldmaking, calling Latour and Woolgar irrealists "at large in the laboratory." However, it seems important to stress here (as above in the Introduction) – since these positions, of Latour and Woolgar on the one hand and Goodman on the other, stood against realism – that what they objected to was "the ready-made world as a touchstone" (Goodman 1984, 29), *not* the reality of things. The irrealist holds a many-worlds-if-any ontology with no commitment to an uninterpreted reality independent of thought, language, or other cultural expression. Instead, reality comes about in a dynamic and historically determinative process, constrained or restrained by the phenomena. In this way, Goodman can say worlds are not made from pure fantasy (1978, 97) but are the product of the refraction of conception into perception.

For describing the things that belong to the world – its phenomena, its parts, its appearance, and its properties – perception and empirical experience are still significant, informed though they certainly are by conception, and conception by history and culture. It is not the phenomena but the "firm foundation" of the world that is lost in Goodman's irrealism, in which "world" becomes "worlds that are but versions" (Goodman 1978, 7). The realist claim that the world-in-itself provides the solid underpinning of our sense of "what there is," and is supposed to be independent of anyone's thought or experience of it, is what Goodman called a "world well lost" (Goodman 1978, 4 and note 3, citing Rorty 1972).

Israel Scheffler (1980, 201–3) raised questions many of us would about Goodman's world-versions and their relation to the concept of "world." In a discussion of *Ways of Worldmaking* as well as of some other representative works of his teacher Nelson Goodman, Scheffler objected (Scheffler 1980, 201) that Goodman's usage of the word "world" was not sufficiently well defined. He structured his critique around the charge that two conflicting interpretations of the word seem to be supported by the work. The first he called the versional interpretation and the second the objectual interpretation, in which "a world is a realm of things (versions or nonversions) referred to or described by a right world-version" (Scheffler 1980, 201). Scheffler noted:

> "World" in Goodman's usage was ambiguous, at times applying to what he called "right world versions," at times applying to the things referred to by such versions ... his claim that we make worlds can be true only for the "versional" but not the "objectual" interpretation of "worlds." Goodman had insisted that, although we distinguish between versions that do and those that do not refer, the things referred to, "and even the stuff they are made of – matter, anti-matter, mind, energy, or whatnot – are fashioned along with the versions themselves" [Goodman 1978, 96]. My attitude from the outset was that "worldmaking" as a metaphor for version-making I could tolerate but "worldmaking" as a literal claim to the making of the objects of such versions I could not. I thought it obvious that we made the word "star," and equally obvious that we did not make the stars, concluding that, despite his disclaimer [Goodman 1978, 110], Goodman's statement "we make worlds by making versions" [1978, 94] must of course be taken as "purely rhetorical." (Scheffler 2001, 668)

Worldmaking with versions, however, was not meant as mere rhetoric or metaphor, as Goodman explicitly said in answer to Scheffler, nor were worlds and versions the same (Goodman 1984, 34; 1996a, 144). By versions he meant symbol systems (verbal descriptions, visual representations) without which one cannot refer to or describe a world. In Ricoeur's review of *Ways of Worldmaking*, he said Goodman's world-version was not "a version of the world," and in order to understand Goodman's notion "we must drop the idea that versions are versions of one and the same neutral and underlying world" (Ricoeur 1980, 108). There is, as a consequence, no description or representation possible of a version-independent (or framework independent) reality. Accordingly, there are no facts independent of versions; a right version is true of its world, thus conflicting right versions are true only of different worlds, and one world cannot sustain conflicting truths.

Science was but one of the ways we make worlds that are true to our own versions (Goodman 1984, 30–39). The fruitfulness of Goodman's line of reasoning for the historiography of science is clear in *Ways of Worldmaking*, where his point of departure was the claim dealt with by Kuhn as a matter of paradigms. Thus he stated:

> Consider, to begin with, the statements "The sun always moves" and "The sun never moves" which, though equally true, are at odds with each other. Shall we say, then, that they describe different worlds, and indeed that there are as many different worlds as there are such mutually exclusive truths? Rather, we are inclined to regard the two strings of words not as complete statements with truth-

values of their own but as elliptical for some such statements as "Under frame of reference A, the sun always moves" and "Under frame of reference B, the sun never moves" – statements that may both be true of the same world.

Frames of reference, though, seem to belong less to what is described than to systems of description: and each of the two statements relates what is described to such a system. If I ask about the world, you can offer to tell me how it is under one or more frames of reference; but *if I insist that you tell me how it is apart from all frames, what can you say?* We are confined to ways of describing whatever is described. (Goodman 1978, 2–3; my emphasis)

The emphasis on description of phenomena is found throughout *Ways of Worldmaking*, as in "if we say that the geocentric and heliocentric systems are different versions of 'the same facts,' we must ask not what these facts are but rather how such phrases as 'version of the same facts' or 'descriptions of the same world' are to be understood" (1978, 93). Facts apart from versions or descriptions of course did not exist in Goodman's formulation.

I find these ideas extremely useful from a historical perspective because they can free our categories and judgments from expectations stemming from assumptions about an "underlying solid foundation of matter" as a universal underpinning for inquiry about unfamiliar systems of worldmaking and their cosmologies. For the historian, the "ways of describing" are available in textual sources, and Goodman was prepared to see in these ways the versions that constitute for their inhabitants actual worlds: "The multiple worlds I countenance are just the actual worlds made by and answering to true or right versions. Worlds possible or impossible supposedly answering to false versions have no place in my philosophy" (Goodman 1978, 94). In response to Scheffler's refusal to believe anything but a metaphorical sense of world-version making,[10] or a rhetorical sense of the whole idea of making worlds by cognitive activity (Scheffler 2009, 67), Goodman replied:

The worldmaking mainly in question here is making not with hands but with minds, or rather with languages or other symbol systems. Yet when I say that worlds are made, *I mean it literally*; and what I mean should be clear from what I have already said. Surely we make versions, and right versions make worlds. And however distinct worlds may be from right versions, making right versions is making worlds. This is a conspicuous case of how *talk of worlds and talk of versions coalesce*. (1996a, 144; my emphasis)

[10] I note that Donald Davidson too asserted that because there is "at most" only one world, any talk of pluralities is metaphorical, purely imaginary (1973, 8–9).

Goodman's ontological evanescence is one of versions, but those versions are not mere worldviews, but actual worlds to which a version refers. The image of a world "melting into versions and versions making worlds" (Goodman 1984, 29), as he put it, seems a rather poetic way of defining the nature of their distinction while preserving the representational nature of versions.[11]

Consistent with his constructivism and pluralism, Goodman rejected the strict dichotomy between world and version, or what monists (whether physicalists or phenomenalists) would see as world and worldview, along with the absolutism of a given independent reality. But the melting of versions into worlds, he admitted, made for an "elusive" distinction between the two (Goodman 1984, 30). What makes Goodman's idea interesting is that he maintained the distinction while insisting on the complete dependence of worlds upon versions, thus repudiating the one-world-many-worldviews model.

The coalescence (melting) of versions and worlds has obvious resonance with Kuhn's paradigms, which he too wished to differentiate from mere difference in interpretation of "the same" phenomena. Both Goodman and Kuhn gave up the idea of a world independent of a version in the sense that it is impossible to say what such an independent world would be, suggesting that versions are built both upon and into the phenomena of our experience. Kuhn asked: "Do we . . . really need to describe what separates Galileo from Aristotle, or Lavoisier from Priestley, as a transformation of vision? Did these men really see different things when looking at the same sorts of objects? Is there any legitimate sense in which we can say that they pursued their research in different worlds?" ([1962] 1996, 120). Addressing the implications and anticipating his critics, Kuhn clarified his sense of different paradigms referring to different worlds in the following way:

> Many readers will surely want to say that what changes with a paradigm is only the scientist's interpretation of observations that themselves are fixed once and for all by the nature of the environment and of the perceptual apparatus. On this view, Priestley and Lavoisier both saw oxygen, but they interpreted their observations differently; Aristotle and Galileo both saw pendulums, but they differed in their interpretations of what they both had seen. (Kuhn [1962] 1996, 120–21)

[11] It is worth mentioning in this context that Goodman preserved a function for representation, indeed that world-versions were the product of symbol systems and representations. This distinguishes his thought from that of the more recent anthropological ontologists mentioned before who disavow representation and see the collapse of symbol and object. See also note 6.

He concluded, however, not that these were mere differences in interpretation but rather that "the proponents of competing paradigms practice their trades in different worlds" (Kuhn 1996, 150; see discussion of Kuhn's concept of world in Hoyningen-Huene 1990, 484–86).

The worldmaking of Goodman rests on his rejection of the existence of autonomous objects in a "fixed and found world" (Goodman 1978, x), or "readymade world" (Goodman 1978, 94, 132 note 20; Goodman in McCormick 1996, 167; also Putnam in McCormick 1996, 190 refers to the "myth of the readymade world") to express his rejection of the idea of truth as correspondence to such a fixed-and-readymade world. In the context of science, of course, the "fixed and found" or "readymade" world is nature. In *Mind and Other Matters* (1984), as in *Ways of Worldmaking* (1978), Goodman stated that "the forms and the laws in our worlds do not lie there ready-made to be discovered but are imposed by world-versions we contrive – in the sciences, the arts, perception, and everyday practice" (Goodman 1984, 21).[12]

Goodman denied the givenness of an antecedent reality and clarified the difference between his view and that of the idealist, who would hold the position he typified as that of "the mystic" (Goodman 1996b, 4). Accordingly, the mystic sees the world filtered through perceptions, cognition, concepts, and language. All these filters produce distorted versions of a single reality, and none of them are quite right by reason of our failure to ever be able to access that single ultimate reality. In his words:

> The mystic holds that there is some way the world is and that this way is not captured by any description. For me, there is no way that is the way the world is; and so of course no description can capture it. But there are many ways the world is, and every true description captures one of them. The difference between my friend[13]

[12] Putnam 1982's rejection of the ready-made world was more a fight with metaphysical realism and with "metaphysical materialism," which he defined (p.147) as *natural* metaphysics. He said (p. 147; emphasis in the original): "What makes the metaphysical realist a *metaphysical* realist is his belief that there is somewhere, One true theory (two theories which are true and complete descriptions of The World would be mere notational variants of each other). In company with a correspondence theory of truth, this belief in One true theory requires a ready-made world – the world itself has to have a 'built-in' structure since otherwise theories with different structures might correctly 'copy' the world (from different perspectives) and truth would lose its Absolute (non-perspectival) character. Moreover, as I already remarked, 'correspondence' between our symbols and something which has no determinate structure is hardly a well-defined notion."

[13] In Goodman 1960, 49, he called the mystic "my arch enemy," who would say, "Science, language, perception, philosophy – none of these can ever be utterly faithful to the world as it is."

[the mystic] and me is, in sum, the enormous difference between absolutism and relativism. (Goodman 1996b, 10)

Goodman rejected absolutism, and with it the single version-independent world as the determinant of the one way the world is.

Goodman's opposition to the mystic would not apply to Descola, in spite of the latter's use of the concept of ontological filters, because, in spite of upholding one-worldism, Descola nowhere claims that there is only one way that the world is, but that its potentialities are realized as differently perceived and understood by human beings as a product of culture. Human beings perceive and understand the potential world and its versions, so to speak, in multiple ways, such as are illustrated by his fourfold ontological scheme, each one formed after "the potential qualities and relations afforded to human cognition ... have been detected and actualized" (Descola 2014, 277).

The principal reason ontological relativism is difficult to comprehend much less accept is because it dissolves what Goodman referred to as "something stolid underneath" (Goodman 1978, 96), the thing for which we all seem to have an intuitive trust, if not need. Kuhn, as well, made sure to account for "something stolid underneath" when he qualified his statement that worlds change with paradigm shifts, by saying "the world ... is not ... fixed once and for all by the nature of the environment, on the one hand, and of science, on the other. Rather, it is determined jointly by the environment and the particular normal-scientific tradition" (Kuhn [1962] 1996, 112). This comes quite close to Goodman's ontological evanescence, where the versions of worlds *make* the actual worlds we inhabit, not from nothing, but through our mental and physical engagement with phenomena. The evanescence of the world, in Goodman's formulation, is tied to the idea that even if we claim as a part of our world construction that there *is* an uninterpreted reality out there somewhere, it would be a matter of convention to say so.

In the following passage, Kuhn, in the 1990 address "The Road since Structure" articulates a position touching many points similar to those of Goodman's worldmaking:

Underlying all these processes of differentiation and change, there must, of course, be something permanent, fixed, and stable. But, like Kant's *Ding an sich*, it is ineffable, undescribable, undiscussible. Located outside of space and time, this Kantian source of stability is the whole from which have been fabricated both creatures and their niches, both the "internal" and the "external" worlds. Experience and description are possible only with the described and describer separated, and the lexical structure which marks that separation can do so in different ways, each resulting in a different, though never wholly different, form of life. Some ways are better suited to some purposes, some to others. But none is to be accepted as true or rejected as false; none gives privileged access to a real, as against an invented, world. (Kuhn [1991] 2000a, 104)

Worldmaking, similarly, does not privilege one world-version over another. Many world-versions are "right," in Goodman's sense, and whatever is stolid underneath loses its hold on our ontological claims along with its supposed mind-independence and status as a preconceptualized and thus uninterpreted reality.

In another context, Kuhn, in the 1991 lecture "The Trouble with the Historical Philosophy of Science," remarked on the mind-independence of the one-worldist view of scientific inquiry in history as in the present:

What replaces the one big mind-independent world about which scientists were once said to discover the truth is the variety of niches within which the practitioners of these various specialties practice their trade. Those niches, which both create and are created by the conceptual and instrumental tools with which their inhabitants practice upon them, are as solid, real, resistant to arbitrary change as the external world was once said to be. But, unlike the so-called external world, they are not independent of mind and culture, and they do not sum to a single coherent whole of which we and the practitioners of all the individual specialties are inhabitants. (Kuhn in Conant and Haughland 2000, 120)

A noteworthy consistency and resonance between Kuhn and Goodman fall principally in their shared pluralism and rejection of one-worldism. Whether we regard the similarities as surface or deeper,[14] Goodman had no historical agenda as did Kuhn. Insofar as each one offered an alternative to one-worldism, I see the importance of both to the historiography of science.

[14] Alexandre Declos says, "Goodman and Kuhn simply claim different things. The latter seems to endorse a *diachronic* view on the plurality of worlds, whose model would roughly be the following: there exists a world W_1 which corresponds to paradigm P_1 at time t, a world W_2 corresponding to paradigm P_2 at t_1^{sic}, etc. Goodman, for his part, develops a *synchronic* model for the plurality of worlds. At one and the same time t, there would exist several actual worlds $W_1, W_2, \ldots, W_n$, with *n* being the total number of conflicting true versions of reality" (2019, 2). The diachronic/synchronic distinction between the claims of Kuhn and Goodman is certainly the case, but the closely related way each argues for the plurality of worlds and the relation those worlds have to the putative and mythical uninterpreted reality suggest to me something more closely related than one being diachronic and the other synchronic.

In my view, the danger in adopting one-worldism as a historiographical foundation is, as Catherine Elgin put it, that "we are forced, if we are fundamentalists, to deny that more than one version can yield understanding. Unless they are reducible to the one right version, and satisfy the criteria of the one right metaversion, the fundamentalist considers works of history, psychology, painting, and philosophy to be cognitively bankrupt" (Elgin 1984, 44). This is precisely why I see the pluralism and irrealism of worldmaking as providing a compelling argument against the notion that the cuneiform sciences were only a beginning step on the road to discovery of the "one true (or right) meta-version" of the real world. The opposite is indeed the case. The character of cuneiform knowledge, the specific ideas that shaped it, and the phenomena defined by it made a world that was not by definition rudimentary or undeveloped in comparison to later science, but legitimate and "right" in the terms of cuneiform culture.

PLURIWORLDISM AND HISTORY

As just set forth, there is historiographical significance in the question of whether or not there is a universal reality to which all minds and languages refer – the one-world-many-worldviews problem – because if there is but one way the world is, then the many historical cultures that have articulated different perspectives on the composition and organization of the world order can only be seen as having misconstrued and misconceived that order. This includes premodern Western cultures with their posits of world stuff such as aether, celestial spheres, caloric, and phlogiston, none of which we now regard as ever having had actual ontological status. In the history of science, we gain very little by entering into an interpretive program about sources stemming from distant times and places with an a priori assumption that their views are misunderstandings of a reality defined by the standards of nearer times and places. Putnam was interested to show that indeed no evidence from the history of science supports the idea "that it [science] either aims at or should aim at one single absolute version of 'the world'" (Putnam 1982, 164). He found such a claim to run counter to the very essence of science, "whose strategy from the first has been to confine itself to claims with clear empirical significance" (Putnam 1982, 164).

Kuhn's question, "Can a world that alters with time and from one community to the next correspond to what is generally referred to as 'the real world'?" and his answer, "I do not see how its right to that title can be denied" (Kuhn in Conant and Haugland 2000, 102) addresses pluriworldism from the question of the legitimacy of ontological posits outmoded by current standards of ontological reality. Kuhn extended to historical communities "rights" to a real world no matter the historical context for that reality. This clearly stems from Kuhn's historiographical, as opposed to Goodman's philosophical, stakes in the matter of the plurality of worlds. Kuhn's many worlds are the consequence of historical change, whereas Goodman's stem from his rejection of absolutism in any form, especially that of materialists or physicalists who reduce all ontology to a single order of reality. The ontological question of what exists in Goodman's thought, therefore, is not solely a matter of historical evidence.

One of the reasons for my interest in ontological pluralism consequent on pluriworldism, which is one of the two principal ideas of Goodman's worldmaking (the other being irrealism), is that Goodman did not assign to any world-version a position of privilege. Accordingly, it seems to me necessary from a historiographical standpoint to establish the validity of alternative world-versions as actual worlds if we are not to fall into hierarchies based on developmental histories, where we note, for example, progress from curious worldviews (Babylonian) to empirically or rationally justified cosmologies (Greek and other premodern Western cosmologies).

Goodman posited a plurality of actual worlds and an ontological relativity, where the version and the world are not conflated and identical but inextricably bound, such that versions denote worlds and worlds answer to right versions. In that way, and in spite of his sometimes speaking as though versions *are* worlds, we can say with Alexandre Declos that "[worlds] are ontological realms, which can be understood as the denotata of versions, or alternatively, as what they are committed to. The latter [versions] are complexes of symbols functioning referentially" (Declos 2019, 7).

Multiple worlds, according to Goodman, are logically necessary in order to answer to multiple versions committed to ontologies capable of supporting conflicting truths. Replying to charges of a mere rhetorical force of his idea of multiple worlds and conflicting truths, Goodman said:

> To anyone but an arrant absolutist, alternative ostensibly conflicting versions often present good and equal claims to truth. We can hardly take conflicting statements as true in the same world without admitting all statements whatsoever

(since all follow from any contradiction) as true in the same world, and that world itself as impossible. Thus we must either reject one of two ostensibly conflicting versions as false, or take them as true in different worlds, or find if we can another way of reconciling them. (Goodman 1978, 110)

One of the confounding implications of Goodman's many worlds was the logical conundrum of the word "world." In terms of its meaning *all there is*, a totality, and therefore logically and necessarily *one*, there cannot be many totalities (Goodman 1978, 32). Declos (2019, 5–6) discussed Bas van Fraassen's (1995, 153) way of considering how the word "world," which is not a "count noun," could indeed be context dependent and therefore plural. Goodman's assertion was:

We usually think and work in one world-version at a time . . . but we shift from one to another often. When we undertake to relate different versions, we introduce multiple worlds. When that becomes awkward, we drop the worlds for the time being and consider only the versions. We are monists, pluralists, or nihilists not quite as the wind blows but as befits the context. (Goodman 1996b, 153)

Much as a multitude of worlds is difficult to accept for the reasons just stated, so too is it difficult to pin down what exactly is entailed by the expression the world *as a whole*. In reflecting on the question of what "the external world as a whole" means, Chang offered that:

it is really not clear what it is that we philosophers have in mind when we talk about "the world." If you think that is a straightforward matter, tell me this: does the world include God in it? Even when cosmologists theorize about the universe, they deal with only particular aspects of the universe conceptualized in very specific ways, not just "all that there is." (Chang 2022, 128)

It is in this context that Chang also refers to Goodman's world-versions as "all that we can ever actually deal with" (Chang 2022, 128). But this partial coming to terms with aspects of the world as we engage with it – where "engage" covers activities such as perceiving, conceiving, and describing (with words, images, and numbers) – either now or in the past or in other cultures, is how knowledge develops within the many frameworks producing and defining the many ways the world is or is construed to be. The inevitable plurality of ways of worldmaking is its outcome. When a pluralist/irrealist uses the word "world," this is what she means.

As already discussed, Goodman's argument for the plurality of worlds did not come primarily from historical evidence (as it did for Kuhn) or out of a concern with historiographical methodology. Rather it was developed as a philosophical argument stemming from the existence of versions with equal claims to truth. Conflicting versions could refer to worlds with different entities, worlds with the same entities but differently ordered, or worlds with different causalities. His examples came both from history, as in the conflicting true versions of geocentrism (the earth stands still) and heliocentrism (the earth moves), and modern mathematics, as in a world where points are seen as the result of pairs of intersecting lines or another where lines consist of points (Goodman 1978, 12–13, 117). In each case, whether the motion or stillness of the earth or the composition of lines and points, they are conventions, not facts, and in the conflicting versions described in each case there is a world answering to them.[15] Conflicting truths in one world are impossible (Goodman 1978, 110), or lead to an unconstrained relativism that Goodman did not defend.

Quite some years after the initial critical responses to Goodman's pluriworldism – mainly by Scheffler, mentioned before, but also in the volume *Starmaking* (McCormick 1996) – a recent exposition, including discussion of its limitations, ambiguities, and replies to its critics, can be found in "Goodman's Many Worlds" of Declos (2019). Declos said:

> Although pluriworldism and worldmaking are entwined in Goodman's writings, it should be noted that they are, at least in principle, logically independent. Many admit that we somehow "make the world," but object to the pluralistic claim that there is more than one world – this is indeed the standard "constructivist" stance (Hacking 1999). Conversely, one could in theory accept that there exist several worlds while maintaining that those are not made but rather found or "ready-made" (see Scheffler 2000). (Declos 2019, 1)

If indeed there are irreconcilable and irreducible world-versions, whether of different entities, of parts, or in the manner of their organization and relationships, those conflicting true versions are only possible in different worlds. This is also why different schemes (or models) do not refer to "the same facts" (for the notion of facts in an uninterpreted reality, see the next section).

In defining and defending Goodman's pluriworldism, Catherine Elgin explained:

> The plurality of acceptable versions results from the existence of mutually irreducible conceptual schemes. Why don't we just take the schemes to provide different characterizations of the same facts? In that case, there are several divergent

[15] The following remark of Putnam is both apposite and amusing: "Metaphysical realists to this day continue to argue about whether points (space–time points, nowadays, rather than points in the plane or in three-dimensional space) are individuals or properties, particulars or mere limits, and so on. My view is that God himself, if he consented to answer the question 'Do points really exist or are they mere limits?' would say, 'I don't know'; not because his omniscience is limited, but because there is a limit to how far questions make sense" ([1987] 1989, 175).

> versions of the same world. This is hardly cause for alarm. The trouble is that we lack the resources to justify, or even make sense of the claim that all overlapping acceptable versions pertain to the same things. Without the scheme/content distinction,[16] there is no way to distinguish the conceptual from the factual ... Since mutually irreducible schemes do not invoke equivalent criteria, they do not treat of the same things. A world is the totality of things that comprise it, so mutually irreducible schemes define distinct worlds. There are, Goodman concludes, many worlds if any. (Elgin 2000, 12)

The remark at the end of the last sentence, "There are ... many worlds if any," refers to Goodman's insistence that one-world monism, many worlds pluralism, realism, and idealism are all conventional, and thus "the world" is a function of constructional systems, languages, and domains of discourse, and (I would add) historical contexts, impossible to be conceived of independently of any one of those conventions. Goodman's own characterization of his position was one of "radical relativism under rigorous restraints" (Goodman 1978, x, 94).[17]

Fending off charges of a relativism of the anything goes and all things are equally valid variety, Goodman said, "ontological relativism does not imply that all world-versions are right but only that at least some irreconcilable versions are right. Examples abound, and a single world can hardly answer to conflicting versions" (Goodman 1984, 9). What he referred to were the many examples of conflicting representations, each warranted individually but impossible to accept simultaneously, as in the well-known representation of light as particles or waves, discussed in light of Goodman's idea of versions by Elgin (1984, 36).

For such divergent, and conflicting, descriptions, as Elgin pointed out, only the "highest cognitive or epistemic standards" are satisfied, and "Goodman's pluralism consists in his recognitions that more than one version does so" (Elgin 1984, 37). As for constructed worlds and fabricated facts, it does not follow, Elgin continued, that "we can construct any worlds and fabricate any facts we please. We cannot, for example, construct a world in which metamorphosis transforms

[16] Elgin refers here to scheme–content dualism, or the "third dogma of empiricism" of Donald Davidson, for which see the next section and Davidson 1973, 11; 2005, 48.

[17] Cladis (1991, 142–43) addressed the question of what restraints, or constraints, on worldmaking Goodman maintained and considered whether this was a mediation between world-finding and world-making, that is, what he, Cladis, called "mild-mannered pragmatism." I fully concur with Cladis's reading of this dynamic, which is also found in Goodman's usage of convention in defining irrealism, for which see above, p. 116.

a caterpillar into a crocodile. Why not? The answer is, simply, because it doesn't" (Elgin 1984, 38). The advantage for the historiography of science lies in the defense of the commitment to the truth of different world-versions, that is to say, the individual warrants for apparently conflicting world-versions for the inhabitants of the worlds answering to those versions.

CONCEPTUAL RELATIVISM AND TRANSLATION

In the opening statement of his 1973 article "On the Very Idea of a Conceptual Scheme," Donald Davidson set down the positions he opposed, positions that underlie conceptual relativism and ontological pluralism:

> Philosophers of many persuasions are prone to talk of conceptual schemes. Conceptual schemes, we are told, are ways of organizing experience; they are systems of categories that give form to the data of sensation; they are points of view from which individuals, cultures, or periods survey the passing scene. There may be no translating from one scheme to another, in which case the beliefs, desires, hopes and bits of knowledge that characterize one person have no true counterparts for the subscriber to another scheme. Reality itself is relative to a scheme: what counts as real in one system may not in another. (Davidson 1973, 5)

This paper was called "one of the most discussed papers in recent philosophy" and "a head-on attack on the relativist tradition" (Blackburn [2005] 2010, 165). Davidson's objection to conceptual schemes did not apply to the association between languages and conceptual schemes, only to the claim that differences cannot be mutually intelligible through translation, intuition, or by whatever means, or "that someone can take up a vantage point for comparing conceptual schemes by temporarily shedding his own" (Davidson 1973, 7). The "very idea" of the conceptual scheme, according to Davidson, was unintelligible if defined in terms of an untranslatable language. As he said, "the 'criterion of languagehood' is translatability or nothing" (Davidson 1973, 14).

That incommensurability was impossible by reason of mutual intelligibility was related to the other important focus of his attack, which was on what he called scheme–content dualism, or the "third dogma of empiricism" (adding to the two dogmas famously called out by Quine 1951 as those of the analytic/synthetic distinction and of reductionism). The third dogma, Davidson said, made a false distinction between an uninterpreted unconceptualized mind-independent given, supplied by empirical or sensory experience of nature on the one hand and the conceptual schemes devised to construct and interpret that given (that is, nature) through language on the other (Davidson 1973, 11; 2005, 48). For Davidson, the scheme *was* language, "with its built-in ontology and theory of the world" (Davidson 2005, 48). Removing the distinction between scheme and content removed any criteria by which to defend conceptual relativism; it also, and more significantly, undermined the role of empiricism itself in science.

Davidson's famous essay has long been looked to as an argument against incomparability, incommensurability, and conceptual relativism. Davidson opposed conceptual relativism with its dominant metaphor of different points of view because of the underlying paradox to which his paper called attention. In his words, "different points of view make sense, but only if there is a common coordinate system on which to plot them; yet the existence of a common system belies the claim of dramatic incomparability" (Davidson 1973, 6). Davidson's disavowal of the putative common coordinate system, the "neutral ground" necessary to differentiate and/or arbitrate between conceptual schemes, has a certain resonance with Goodman's disavowed "uninterpreted reality," although to different ends: absolutism for Davidson, pluralism for Goodman. As Davidson said:

> It would be wrong to summarize by saying we have shown how communication is possible between people who have different schemes, a way that works without need of what there cannot be, namely a neutral ground, or a common coordinate system. For we have found no intelligible basis on which it can be said that schemes are different. It would be equally wrong to announce the glorious news that all mankind – all speakers of language, at least – share a common scheme and ontology. For if we cannot intelligibly say that schemes are different, neither can we intelligibly say that they are one. (Davidson 1973, 20)

Alternative conceptual schemes, that is, the differentiation between them, in Davidson's formulation depended upon the existence of untranslatable languages, an idea he found impossible and indefensible. Davidson's claim that there is "no such thing" as an untranslatable language was not a denial of difference, even of ontological difference between languages; it was a claim only that they did not represent the sort of conflicting truths discussed by Goodman.

It was the denial of conflict between versions, to use Goodman's term that distinguished Davidson's claim. In his words: "Of course, if the 'alternative ontologies' or 'envisaged worlds' were somehow in conflict, we could not accept them simultaneously. Since the various worlds and ontologies are generated by equally justified, though different, systems of translation from an alien tongue into our home language, the issue is not whether we can describe them all. But do they conflict?" (Davidson 2005, 76). In spite of his reference to ontologies and worlds, Davidson, both in "The Very Idea" and in his 2005 *Truth, Language and History*, was mostly in dialogue with Quine not Goodman, and the matter of difference and relativism was delimited by the problem of translation.

In *The Structure of Scientific Revolutions*, Kuhn did not use the term "conceptual scheme"; however, his remarks on "communication breakdowns" faced by those adhering to different "paradigms" was a preview of the problems brought out by Davidson's "The Very Idea." Kuhn attributed the communication breakdown between members of different paradigm-worlds to the fact that "they cannot … resort to a neutral language which both use in the same way and which is adequate to the statement of both their theories or even of both those theories' empirical consequences" (Kuhn [1962] 1996, 201). His answer to the situation of conflicting schemes, paradigms, or "versions" is that their participants would become translators: each "will have learned enough to translate the other's theory and its consequences into his own language and simultaneously to describe in his language the world to which that theory applies. That is what the historian of science regularly does (or should) when dealing with out-of-date scientific theories" (Kuhn [1962] 1996, 202).

At this juncture, it is important to note the relevance of the foregoing debate to contemporary anthropology and thus to close this chapter by bringing the concerns of anthropologists back into focus. The role of translation in negotiating the worlds of others was the focus of a special issue of *HAU: Journal of Ethnographic Theory* (vol. 4) on "Translating Worlds." In their introduction to this issue, William Hanks and Carlo Severi noted, "there is more to translation than language" (2014, 7). Davidson's criticism of Quine was not mentioned in this context, but Quine was not taken on faith either. In a departure from Quine's inscrutability of reference and indeterminacy of translation and an embrace of Kuhn, Hanks and Severi expressed a renewed interest in translation for anthropology as an epistemological principle for the anthropological project of the understanding of other worlds.

The role of translation as an anthropological method, then, is fundamental for all research into other cultures. Although the pragmatics and semantics of "live" multilinguistic encounters is not matched in today's methods for study of the dead cuneiform languages within Assyriology as compared against field ethnography, the problem of reference and meaning of words certainly is a factor for understanding the worlds reflected in dead languages too. At the same time, Hanks and Severi argued that although translation is the necessary starting point, it is not the exclusive domain of understanding, as other aspects deriving from the contexts of communication serve as additional frames for meaning. Certain kinds of nonverbal contexts of communication may be unavailable to the Assyriologist, but certain other nonverbal media, such as are found in representation, in art (imagery) as in science (numerical description, schemata and modeling), provide other domains of understanding. In so far as art and science have the capacity to denote and refer to worlds and their phenomena, they can play a supplementary role in translating between worlds.

Indeed, context is not uniquely important to field ethnography but also to research in ancient scientific (or any) texts. In the case of cuneiform scientific texts, we do in fact rely on broader cultural contexts for the epistemic values, theories, and ontological commitments that may not be explicit in those texts but are the essence of worldmaking. Hanks and Severi remarked that "the interpretation of such formal contexts of cultural representation transforms translation into a way to translate '*worlds*' (defined as 'oriented contexts for the apprehension of reality'), not just words, or other ways to make meaning" (Hanks and Severi 2014, 8; emphasis in the original). The description and representation of worlds are certainly not limited to words but include numbers, images, and symbols, every conceivable mode of expression of worldmaking.

The purpose of this chapter was to discuss the major issues attending the terms worlds, world-versions, and worldviews that surround worldmaking with its derived pluriworldism and multiple ontologies, and of the interconnecting threads from philosophy to anthropology to the history of science. My aim was to establish a framework for situating a discussion of cuneiform science as a way of worldmaking in the remaining chapters and to bring our consideration of it within bounds of an anthropology of ancient science.

Some will still see cultural and historical representations and descriptions of worlds as evidence merely of other worldviews and not of actual other worlds. However, should we remain committed to the existence of a single autonomous world, knowledge in the past can only be adjudged as true because it corresponds to or approaches current standards of truth about it or false because it does not. Judgment of historical systems of knowledge and their accompanying methods and practices on that basis elides important distinctions between the ways that science serves to resolve the all-important relation between the perceived and experienced world on the one hand and human thought in different places and different times on the other.

The historiographical implications of the ideas of the philosophers, anthropologists, and sociologists of science discussed in this chapter are developed in the next chapters by looking at the uniqueness of cuneiform classification and model-making, as well as persistent themes in the organization of cosmological world parts attested in the cuneiform texts.

5

Uses of Diversity

But we are all human, I thought, wondering what I meant.
– Ralph Ellison, *Invisible Man*

It may seem a curious exercise to reflect on the question of diversity or ethnocentrism in the context of ancient historical texts and cultures. The reason for this chapter, however, is to make the point that, although the historiography of science has entered a period of pronounced counter-ethnocentrism since the 1960s and 1970s, there is still much to do to embrace its potential for the science of the ancient Middle East more fully.[1] Here, I offer a close look at some aspects of cuneiform classification and modeling of astronomical phenomena as a basis on which to reflect. In addition, this chapter begins to address more directly how a central theme for the anthropology of ancient science, namely, worldmaking, relates to cuneiform science.

The assumptions we bring to our interpretation of the historical ontologies of our sources underlie the interpretive divide between, on the one hand, the idea that a multiplicity of worldviews are "takes" on a single world, where we (with the benefit of modern science) become arbiters of what that world is, and, on the other, the multiple actual worlds of culturally constructed systems, with their (to us) "exorbitant ontological

[1] In 2001, CHAMA (Commission for the History of Ancient and Medieval Astronomy), a commission of the International Union of History and Philosophy of Science (IUHPS), was created at the 21st Congress of the Division of History of Science and Technology to support research in the astronomical heritage of all cultural areas of the world. Science and *cultural diversity* was the stated theme of the 21st Congress of this body (IUHPS) in Mexico City.

cost" (Declos 2019, 13). That debate bears on the subject at hand concerning the uses of diversity and how we negotiate the question of one universal common world to which all constructed versions can (or should) be reduced versus the irreducible differences seen across cultures and historical evidence and the multiple worlds they reference.

The question of how to reconcile cultural diversity and human universality holds particular resonance for anthropology. It is an enduring theme in the work of anthropologist and ethnographer Clifford Geertz as well as historical anthropologist Geoffrey Lloyd. Lloyd in particular directed the question toward science in its diversity of cultural worlds. How we view the diversity of the science reflected in cuneiform texts as evidence of a difference from ways we ourselves view and understand our world is central to an anthropology of ancient science.

AGAINST UNIVERSALISM

When Clifford Geertz delivered his Tanner Lecture on Human Values at the University of Michigan in 1985 with a talk titled "The Uses of Diversity," he spoke of what he saw as a new trend in the form of "the possibility that the variety [of human cultural difference] is rapidly softening into a paler, and narrower, spectrum" (Geertz 1986, 253; also Geertz 2000, chapter 4). We can now see that the diminishment of difference that Geertz remarked upon as the result of a more global world was fleeting and illusory. Geertz saw the UN General Assembly declare 1971 the International Year for Action to Combat Racism and Racial Discrimination and then thirty years later name the year 2001 as the International Year of Mobilization against Racism, Racial Discrimination, Xenophobia and Related Intolerance.

Despite those noble declarations, the world has only seen further calcification of injustice, discrimination, exclusion, and oppression. If 1971 or 2001 had something to offer racial reconciliation or to put an end to bigotry and systemic racism, those efforts have long been undermined and forgotten. Geertz's piece stemmed from reflection on two essays in which Lévi-Strauss ([1983] 1992, 3–36) made plain that ethnocentric language, in the sense of a recognition of difference, is not to be equated with the language of oppressive racism.

Though it proved to have been illusory, Geertz likened the perceived trend in the diminishment of difference to "the desperate tolerance of UNESCO cosmopolitanism" (1986, 257). What was important about this to Geertz was that "it raises a broader issue, moral, aesthetic, and cognitive at once, that is much more troubling, and that lies at the center of much current discussion about how it is that values are to be justified: what I will call, just to have something that sticks in the mind, The Future of Ethnocentrism" (1986, 253). I daresay Geertz would be none too pleased to see the "future" of ethnocentrism so well and thriving, even becoming ever more entrenched in dark and pernicious forms, explicit and implicit, insidious and overt. His essay, however, was no advertisement for ethnocentrism in its dark manifestations but a critique of reductive universalism made possible by cultural anthropological inquiry into multiple orderings of reality.

In the very same year as Geertz's Tanner lecture (1985), Ernest Gellner was equally preoccupied with questions of relativism and the existence of human universals. Gellner defined relativism's core question as "whether all the divergent visions of reality can in the end be shown (leaving out cases when they are simply mistaken) to be diverse aspects of one and the same objective world, whose diversity can itself be explained in terms of the properties or laws of that world" (1985, 87). Gellner's unflinching commitment to the "one and only one world" was not tied to the idea of the single nature of humanity or the existence of human universals. On the contrary, he held that "the uniqueness of the world hinges on the diversity, or non-universality of man" (87). For him "relativism is about the existence of One World" (100), and that one world was both necessary for science ("science needs one world," he said) and was the product only of modern Western scientific thought and not of other cultural systems. The world referred to in Gellner's assertion about science is the fixed-and-found world, to use Goodman's turn of phrase (Gellner 1985, 100). In historical contexts especially, Gellner's view of the superiority of Western science can strike an ethnocentric chord.

An ongoing concern among anthropologists, ethnocentrism is also relevant for the history of science, for, as Jonathan Benthall pointed out, the critique of ethnocentrism is "allied to the precaution against 'presentism' and 'teleology'" (2021, 20) and is dedicated to the decolonization of our canon of knowledge. Benthall's critique of ethnocentrism gives a useful history of the initial coinage and meaning of the term "ethnocentrism," the various figures from the non-Western past who acknowledged human diversity (e.g., al-Bīrūni, who spanned the tenth and eleventh centuries; see further Sturman 2021), and the emergence of counterethnocentrism. The origin of the term is tied to Ludwig Gumplowicz (1879), who, making the analogy to geocentrism, meant the delusional belief in the centrality and superiority of one's own ethnic group. Thereafter in

American sociology (Sumner 1906), the term also carried the belief that the point of view of one's own group was right and true, indeed, that it should serve as the yardstick by which others would be judged (Benthall 2021, 20).

Past scholarship about scribal knowledge in Middle Eastern antiquity once reflected attitudes about its cultures not just of difference but of otherness, that is, a quality of "less than" as compared with Western modernity, or even Western antiquity, meaning Classical and Hellenistic Greek philosophy. This was based on little more than chauvinism for our more immediate intellectual ancestors. The ancient Greeks themselves, however, such as Plato, Herodotus, Aristotle, Pseudo-Plato, and Diodorus (von Staden 1992, 579 note 4; Bowen 2013, 305–6, on Diodorus), placed the origins of mathematics and astronomy in Egypt and generally had a high opinion of knowledge from their eastern intellectual predecessors, the Babylonians included. Even more serious a flaw in historiography is the very idea of the origins of a singular science, an idea that plays into the ethnocentric goal of identifying where our own science began and eliminating everything that didn't participate in that linear intellectual ascent.

When it comes to modern scholarship of a previous era, W. G. Lambert, one of the preeminent cuneiformists of the twentieth century, dismissed the mythological accounts of the birth and genealogy of gods in the Sumero-Babylonian tradition as failed "scientific cosmology," saying:

> In a sense, then, the Babylonian cosmological ideas were a dead end. An overwhelming interest in city theology had deprived them of any desire to pursue cosmology in a scientific manner, and the ideas of the stone age were passed on virtually unchanged . . . The Greeks, by handling the notion of primeval elements impersonally, were able to move on to abstract scientifically biased cosmology. (1975, 55, 61)

Lambert's view of early cuneiform cosmological narratives as a failure to investigate the world as "cosmos" in the ancient Greek way of the cosmologists appears to us as typically ethnocentric. Not only did he discredit the very characteristic (city theology) of such narratives for not aligning with "scientific" (Greek) cosmologies, but he also charged Babylonian ideas about the world as being "a dead end" because they did not lead up to the Classical Western developmental course of cosmology, defined as geocentrism and a finite universe followed after many centuries by heliocentrism and an infinite universe. Against the image of a Greek scientific cosmology, Sumerian and Babylonian descriptions of the world were seen at that stage in our historiography to represent both a failure of conception and the failure of their program of inquiry.

Since that time, both cuneiform and Greek evidence, in some cases separated by many centuries and found in vastly different contexts, has been reassessed and revised (the former in Pingree 1992a and von Staden 1992, the latter in Lehoux 2019). But until such reassessments were made, the entrenched view about "Mesopotamian" thought in comparison to Greek was given widespread long-lasting currency by Henri Frankfort et al. in the *Intellectual Adventure of Ancient Man* (critiqued in Rochberg 2016a, 2016b). There the comparison of Sumerian mythological narratives, rooted in the third millennium BCE, to Greek philosophy of the Classical period served to separate one ancient culture from another in terms of cognitive primitivity or advancement. This is signaled by Lambert's use of "the stone age" as a marker for the ancient Sumerians.

Even Kuhn's *Structure* ([1962] 1996), which articulated the idea of science as the product of communities and a historical phenomenon, did not make room for ancient non-Western scientific cultures. Kuhn's framework for identifying the communities that arbitrated paradigm debates was Europe of the past four centuries. Cuneiform culture, by his reckoning, would have been categorically and intrinsically unable to produce science. In his words:

> Every civilization of which we have records has possessed a technology, an art, a religion, a political system, laws, and so on. In many cases those facets of civilization have been as developed as our own. But only the civilizations that descend from Hellenic Greece have possessed more than the most rudimentary science. The bulk of scientific knowledge is a product of Europe in the last four centuries. No other place and time has supported the very special communities from which scientific productivity comes. ([1962] 1996, 167–68)

Kuhn's scientific communities were defined first and foremost for their concern "to solve problems about the behavior of nature" (Kuhn [1962] 1996, 168) and, second, that the paradigm debates themselves must necessarily be driven by nature itself, "by making prior achievements seem problematic" (Kuhn [1962] 1996, 169). In this statement, Babylonian astronomy would have to have come under the category of technology, or perhaps "rudimentary science." Indeed, if science is confined to solving problems driven by nature, then societies such as those of the cities of Babylon or Uruk in the fifth century BCE, or of Babylon in the eighteenth century BCE did not participate in the history of science.

The interest of the present discussion in the historiography of science is at the same time the reason for bringing the matter within consideration of ethnocentrism. That is, historical analysis, whether because of too narrow a view or because modern criteria are used uncritically for measuring historical ideas, implicitly presumes that it proceeds from a position that is itself not utterly cultural. Nowhere is this position considered more immune to criticism than in the realm of science, our culture's most privileged form of knowledge. Science, in our culture, has acquired immunity from criticism that it might not represent universal knowledge of the way things *are*. But the human cognitive activity we define as science has manifested itself differently over time as a direct consequence of changes in the construal, definition, and understanding of what the world *is* for the people living in it.

Ethnocentrism is purveyed from a mainstream point of view that cannot see itself as the intellectual equivalent of some other point of view somewhere else. The reason for this is not simply the superiority that comes with the ethnocentric point of view, but, as Geertz had it, this "trouble with ethnocentrism" is precisely that we are kept from the realization that we too stand at a particular angle to the world.[2] The same point was foreshadowed in a certain way by Quentin Skinner, when he said that "to recognize ... that our own society is no different from any other in having its own local beliefs and arrangement of social and political life is already to have reached a quite different and ... a very much more salutary point of vantage" (Skinner 1969, 52). Skinner from a sociopolitical historical perspective and Geertz from an anthropological perspective were both concerned to dislodge the vantage point of modern Western values from what was broadly assumed to be the position of ultimate and correct judgment.

GEERTZ ON THE FUTURE OF ETHNOCENTRISM

The aim of Geertz's "Uses of Diversity" was to preserve the value of difference, which, after all, is the stock in trade of anthropology. His point of departure was Lévi-Strauss's claim that racism and ethnocentrism are not (necessarily) one and the same, and that no human society can free itself from the ethnocentric gaze of the rest of humanity (Geertz 1986,

[2] Geertz 1986, 261 alluded to "Forster's Cavafy," which points to a line in E. M. Forster's description of the poet Constantine Peter Cavafy (1863–1933) as "a Greek gentleman in a straw hat, standing absolutely motionless at a slight angle to the universe" (1923, 110).

255–56; Lévi-Strauss [1983] 1992, 24). In fact, to do away with difference, said both Lévi-Strauss and Geertz, would diminish the integrity of individual cultures and thus compromise their diversity. The preservation of diversity, as developed in his essay, is critical to the health of the discipline. As Geertz said, "an anthropology so afraid of destroying cultural integrity and creativity, our own and everyone else's, by drawing near to other people, engaging them, seeking to grasp them in their immediacy and their difference, is destined to perish of an inanition for which no manipulations of objectivized data sets can compensate" (Geertz 1986, 260). Thus, the "trouble with ethnocentrism," as he put it, "is not that it commits us to our own commitments. We are, by definition, so committed, as we are to having our own headaches." It was rather that "it impedes us from discovering at what sort of angle, like Forster's Cavafy, we stand to the world" (Geertz 1986, 261). While this is a central commitment for the discipline of anthropology, it seems to me indispensable, too, for any kind of critical (as opposed to simply descriptive) history of science.

Geertz was equally opposed to the view that there can be no mutual intelligibility between groups, to use Geoffrey Lloyd's terminology:

> The perception that meaning, in the form of interpretable signs – sounds, images, feelings, artifacts, gestures – comes to exist only within language games, communities of discourse, intersubjective systems of reference, ways of worldmaking; that it arises within the frame of concrete social interaction in which something is a something for a you and a me, and not in some secret grotto in the head; and that it is through and through historical, hammered out in the flow of events, is read to imply (as, in my opinion, neither Malinowski nor Wittgenstein – nor for that matter Kuhn or Foucault) meant it to imply – that human communities are, or should be, semantic monads, nearly windowless. (Geertz 1986, 262)

The purpose, the "uses of cultural diversity," lay in its potential for mutual intelligibility. Geertz maintained that only through the study and analysis of difference could we hope to cross the spaces between us, be they geographical, chronological, linguistic, or cultural. Geertz's "uses of diversity" were for us to "know one another, and live with that knowledge, or end marooned in a Beckett-world of colliding soliloquy" (Geertz 1986, 271).

Geertz's deliberately provocative plea for the "future of ethnocentrism" was therefore meant not to defend ethnocentrism but to oppose the "anti-ethnocentrist" overemphasis on human universals and the view that difference, being superficial, is less meaningful than underlying commonalities. Geertz advocated "anti-anti-ethnocentrism," rejecting the combined notions of sociobiology with absolute truth, seeing culture as formative and meaningful.

There is much here in common with Geertz's own 1984 lecture (republished in Geertz 2000, 42–67) titled "Anti-anti-relativism," which did not seek to defend relativism but rather to attack antirelativism with its unfounded fears of such extreme consequences as self-refuting incoherence, a complete collapse of values, and the fall of objectivity to reckless subjectivism. Geertz's critique of antirelativism comes down to his objection to "placing morality beyond culture and knowledge beyond both" (Geertz 1984, 276). In short, Geertz's anti-antirelativist stance was particularly focused on universalism's denial of the force of frameworks for what exists, what is good, or what is true. The negation of absolutisms of various kinds is in fact the central claim of relativism and, contrary to the charges of absolutists, is not itself a self-refuting claim.

In the essay "Anti-anti-relativism," Geertz set the value of diversity against the view that underneath it all human beings are reduced to putative universal characteristics of human nature and the human mind. He observed that "where the Human Nature tack leads to bringing back one of our classical conceptions into the center of our attention – 'social deviance' – the Human Mind tack leads to bringing back another – 'primitive (*sauvage*, primary, preliterate) thought'" (1984, 273). As an example, he pointed to Robin Horton. Paraphrasing from Horton (1982, 260), Geertz said:

> Robin Horton (1982) argues for a "cognitive common core," a culturally universal, only trivially variant, "primary theory" of the world as filled with middle-sized, enduring objects, interrelated in terms of a "push–pull" concept of causality, five spatial dichotomies (left/right, above/below, etc.), a temporal trichotomy (before/at the same time/after) and two categorical distinctions (human/nonhuman, self/other), the existence of which insures that "Relativism is bound to fail whilst Universalism may, some day, succeed." (Geertz 1984, 274)

Horton was centrally concerned about the unities to be found across the diversity of African traditional cultures in the 1982 paper just quoted and also already in his 1967 papers (Horton 1967a and 1967b). He found unity and commonality in what he called "patterns of thought," not just of the cultures of Africa but of "earlier phases of Mediterranean, Near Eastern and European thought" (Horton 1982, 205). This gave a certain evolutionary cast, or even psychic unity flavor, to his observations, namely that traditional cultures are like ancient ones because they have remained in an early phase of human history as compared with modern industrialized people. The other goal of Horton's comparative work was the idea that new light could be shed on traditional African patterns of thought by viewing them through the European modern lens and, as well, that European modern cultural patterns could be reappraised in light of African ones.

In the same year that Geertz's Tanner lecture was published, Richard Rorty published a response, "On Ethnocentrism: A Reply to Clifford Geertz" (Rorty 1986). He too took an anti-antiethnocentric position, delighting in pointing out some ironies, such as that "we would rather die than be ethnocentric, but ethnocentrism is precisely the conviction that one would rather die than share certain beliefs" (Rorty 1986, 525) and, in becoming antiethnocentrists, "we have become so open-minded that our brains have fallen out" (Rorty 1986, 526). Rorty, using Geertz's metaphor of cultures as monads, maintained that our culture is not as windowless as others (Geertz did not view culture in this way but rather critiqued those who *would*, as the above-quoted passage attests). For Rorty, then, anti-antiethnocentrism did not mean we were "trapped within our monad ... but merely that the well-windowed monad we live in is no more closely linked to the nature of humanity or the demands of rationality than the relatively windowless monads which surround us" (Rorty 1986, 526).

Rorty's anti-antiethnocentrism was intended "as a protest against the persistence of Enlightenment rhetoric in an era in which our connoisseurship of diversity has made this rhetoric seem self-deceptive and sterile" (Rorty 1986, 532). His position was political and social, as he opposed the one-world polity with a common culture of the Enlightenment while advocating for a public moral commitment to others, different from ourselves. His claim was that our own ideals and values are as "local" and "culture-bound" as any other, not more representative of an absolute ideal of human nature or of rationality but no less "worth fighting for." The thrust, therefore, was along different lines from Geertz, but both ultimately came down on the side of difference as real, not merely as cultural superstructures upon a universally common human nature.

Differences between the cultures of ancient and other (including modern) sciences are easy to see, but likenesses can also be found if we choose to focus on them that way. How we decide to locate such likenesses as rationality, empiricism, theory, mathematical modeling, logic, and prediction, whether at the deep or the surface level, seems a function of how committed the interpreter is to the deep reality of universals (Brown 1991, 2004) and culture as merely perched upon its surface. For a pluralist, or anti-antirelativist such as Geertz, likenesses across cultures were not evidence of that deep universality, or as Richard Shweder put it: "No 'mind for all cultures.' No 'deep down homo.' No 'dog under the skin'" (Shweder 2000, 1511).

Rorty was interested in the shifting norms of living, or historical, cultures. The normative, even the rational, in the eyes of a given community at a given time varies and changes. From a historicist, postmodern, relativist, antirealist, neopragmatist position, ontology is as subject to cultural and historical variance as epistemology and metaphysics, as the historical and cultural panorama of modes and manners of various human communities' perception, expression, interpretation, and understanding of their *worlds* indicates. In other words, like Geertz and (later) Lloyd, Rorty did not view cultural norms as mere superstructures upon deeper universal things.

Geertz, Rorty, and Lloyd, who is the subject of the following section, in various ways, indirectly in the case of Geertz and Rorty, directly in the case of Lloyd, championed the view that pluralism and "anti-antirelativism" are good for history, especially when analysis and interpretation of other cultures' conceptions of their worlds is at stake. But the question of relativism and universals of human nature, the human mind, and nature itself persists.

G. E. R. LLOYD'S MUTUAL INTELLIGIBILITY

The simultaneity of diversity and commonality has been an abiding concern of Sir Geoffrey Lloyd, whose *Cognitive Variations* (2007), *History and Human Nature* (2010), "Multidimensional Reality" (2011), "The Clash of Ontologies and the Problems of Translation and Mutual Intelligibility" (2019), and *Intelligence and Intelligibility* (2020) offer a wide palette of relevant subjects and approaches to the central issue of mutual intelligibility. Mutual intelligibility may not entirely preclude the pernicious forms of ethnocentrism, where one's own views are sovereign and others' are inferior, but ethnocentrisms of those stripes pose a major impediment to knowing and understanding others.

Lloyd's decolonizing of Greek science is traceable long before these more recent of his works. Lloyd's *Methods and Problems in Greek Science* (1991) and "Methods and Problems in the History of Ancient Science: The Greek Case" (1992) already mounted a historicist and socioanthropological approach to science. In those works, he confronted the problem of ridding ourselves sufficiently of our modern categories in order better to correctly identify and reconstruct the contexts for the inquiry into nature that constituted ancient Greek science. He recognized the pluralism inherent to those contexts and to the Greeks' own categories of thought and practice in the study of nature. Appealing to the anthropological method and to Geertz himself, Lloyd said:

> Although students of ancient societies cannot, in the strict sense, be ethnographers ... that should not prevent us from emulating, as far as our materials permit, the kind of dense – or as Clifford Geertz calls it, "thick" – account that considers the data concerning belief systems against the whole background of the geographical, economic, technological, social and political situation of the society under investigation. (Lloyd 1992, 567)

Following the tour de force *Cognitive Variations: Reflections on the Unity and Diversity of the Human Mind* (2007), Lloyd's "History and Human Nature: Cross-cultural Universals and Cultural Relativities" (2010) began to synthesize a number of themes reaching beyond the bounds of the history and philosophy of ancient science to aspects of human cognition and reasoning (also Lloyd 2020 and Lloyd and Vilaça 2023). Lloyd's 2010 article invited historians, anthropologists, psychologists, linguists, and cognitive scientists to consider evidence of human culture, thought, and reasoning from the standpoint of a difficult duality, namely that the human mind, a dubious abstraction in and of itself, seemingly exhibits both cognitive universalities as well as the capacity for endless diversity of self-expression. As an approach to understanding this *coniunctio*, the article "History and Human Nature" gestured in a direction away from binary conclusions, away from claims that either we are divided by our cultural differences or we are lifted out of our cultures by shared universals. It probed the question of whether cultural identity, as defining, is in fact to be opposed by a state of being merely human.

Coming to terms with both the unity and the diversity of human cultural and mental life over both space and time cannot be reduced to a choice between the positions of the cultural relativist or the cross-cultural universalist, as Lloyd has shown throughout his work. Because cultural diversity *and* shared humanity are each part of what it means to be human on all levels of human existence and experience – social, cultural, intellectual, and individual – whether or how the apparent duality can be reconciled, overcome, or simply better understood is the big question. There will still be debate over which is explanatory, or more explanatory, and what is thereby explained. It is for anthropologists to clarify what separates *us* in the modern West from what has been clumsily and problematically termed *them* everywhere else in an increasingly heterogeneous yet global culture. It is for historians to say what relates the living *us* to the dead *them* of the past, as we no longer view the present in linear or inexorably progressive descent from our forebears. It is for psychologists to explain how each one of us (the individual self) is simultaneously separate from and connected to every other one of us, and it is for philosophers to analyze the extent to which the human mind itself comprehends what it takes to be physical reality, that is, how the world is experienced, perceived, and known. Each of these disciplinary responsibilities is relevant to the historiography of science.

Science is certainly one of the many subjects in which such disparate yet interrelated questions come together. Among philosophers of science, one focus of debate is epistemological, aimed at the relation of mind and world. Objectivity, sense data, and empiricism, the mental, the physical, and the intermediary of language, are not only a crux of the philosophy of mind but also of the philosophy and history of science, where what is at stake is whether any given systematic understanding of physical phenomena is a description of what is really real or, lowering the stakes in truth, simply a way of knowing. A problem for the anthropology of science, as Latour put it, has been that *we* (modern Westerners) have presumed to mobilize nature, setting what *we* do in our sciences apart from what the putative *they* do in their struggles to understand the phenomena of their particular interest:

> We do not mobilize an image or a symbolic representation of Nature, the way the other societies do, but Nature as it is, or at least as it is known to the sciences – which remain in the background, unstudied, unstudiable, miraculously conflated with Nature itself. Thus at the heart of the question of relativism we find the question of science. (Latour 1993, 97)

This thirty-year old critique of the modern scientific establishment might as well have included historians of science who would carry this assumption into historical research, a position common to early historiographies of premodern science, but is steadily fading.

Lloyd was forthright about "the weaknesses of both the cross-cultural universalist and the cultural relativist positions" (Lloyd 2010, also in Lloyd 2006) as well as the "trap" of generalizing about cultures as fixed homogeneous systems, for example, "the Greeks," or "the Chinese," the "semantic monads, nearly windowless" of Geertz (2000, 76), cited before.[3] The objective is not to fix borders around cultures by means of their differences and then to inquire whether within those constructs there

[3] Holbraad and Pedersen's criticism of the "preposterous" nature of such general constructs also comes to mind (2017, 176), discussed in the Introduction.

might still be found deep-level universals. That apparent anomalies are surface and perceptual but conceal underlying uniformities, while sometimes a favored approach to the understanding of certain natural phenomena, does not seem satisfactory for the study of human culture. As Shweder, paraphrasing Geertz, put it, "Differences are not necessarily shallow. Likeness is not necessarily deep" (Shweder 2000, 1511–12, in reference to Geertz 2000, 53).

Lloyd has consistently steered clear of categorical and structural binaries in order instead to analyze the essential tension between diversity and unity/commonality/universality, saying, "We cannot afford to secure uniformity at the price of downplaying diversity. At the same time recognizing that difference should not lead us to lose sight of underlying commonalities" (2020, 2). As indicated in this discussion, Lloyd has done the most to show and explicate the tensions and dynamics within a fusion, or coincidence of opposites (*coincidentia oppositorum*).

A different kind of fusion, not of opposites but of fields of interest and methodologies, also characterizes Lloyd's approach to cross-cultural comparison and the question of translation, or translatability across radically different ways of worldmaking. Together with Aparecida Vilaça, Lloyd discussed ontologies within the cross-cultural and cross-historical context of ancient Greece, ancient China, and contemporary Amerindian (Wari') notions of transformation between human and nonhuman agents, bringing the theme of their dialogue centrally within the anthropologists' ontological turn (Lloyd and Vilaça 2023). Their book, in dialogue form, moves firmly away from cultural representationalism and the idea that the multiplicity of representations amounts to many worldviews about one nature (the external world). It approaches ontologies without having to pass through naturalism or the nature/culture divide as a fixed and universal dichotomy. Because transformation within and between human and nonhuman agencies is their main concern, they consider the analytic and interpretive possibility of an ontology not only of being but of becoming.

An ontology of becoming allows for the "translation" of the views of the Wari' on the jaguar–human and human–jaguar transformations with minimal distortion in Western academic vocabulary and discourse. Lloyd explained:

One major problem is that "ontology" began in philosophy, where it was used to talk about whatever exists as if that should be given a single all-encompassing answer. But that tends to leave no room whatsoever for multiple ontologies and so

> leads quickly to a swift dismissal of whatever does not chime with your own starting assumptions. But if "one-world" solutions suffer from that flaw – we should learn to treat divergent views with more respect – "multiple-world" ones owe the reader some account of their multiplicity. (Lloyd and Vilaça 2023, 96)

Further to that point is Vilaça's observation that, whereas the Western perspective on culture tends to think in terms of "humanity," "as an anthropologist," she said, "I aim at 'humanities' (plural) and fear that ignoring that plurality would lead me to lose the main points that the Indigenous people themselves are trying to teach me" (Lloyd and Vilaça 2023, 107).

The approaches and cooperation between anthropology and philosophy that Lloyd and Vilaça discussed complements and supports my sense that coming to terms with the evidence from non-Western and premodern history of science will benefit from an anthropology of science. In terms of the cuneiform sciences of *ṭupšarrūtu*, in which we find descriptions of phenomena and ways of construing the world unlike those of other non-Western or premodern sciences, the anthropological "uses of diversity" – of cultures, humanities and worlds – can mitigate the ahistorical consequences of assuming either the nature/culture dichotomy or naturalism as an essential for the production of science and eliminate precursorism in our analyses of cuneiform sciences.

I turn now to some examples of how cuneiform evidence affords a view into ontological difference and worldmaking, first in the classification of words and things and second in how model-making can indicate ways of seeing the relation of phenomena to the world.

CLASSIFICATION IN CUNEIFORM

The quasi-Western, quasi-non-Western cuneiform culture of the ancient Middle East, whose celestial sciences are not as widely known as Greek astronomy but which have certain commonalities with that later tradition, provides an opportunity to reflect upon issues concerning likeness, difference, and relativism in the history of science. In terms of its long program of empirical observation and application of quantitative methods to heavenly phenomena, Babylonia represents the foundation and origin of Western astronomy, just as in terms of urbanism, literacy, and statecraft, ancient "Mesopotamia" was once dubbed the cradle of (Western) civilization. But despite holding a position at the Creation (of the West), Babylonian mathematical astronomy in its early reception into the history of science was viewed as a kind of scientific miracle within an otherwise nonscientific belief system consisting of divination and astrology (critiqued in Rochberg 2004, chapter 1). This narrative was based on the tendency to unify when it comes to science and its universal character (i.e., it is about nature not culture) and left other aspects of culture from the cradle to take up a more indigenous character.

This reconstruction was based, however unconsciously, on an ethnocentric historiographical premise, namely, of the scientificity of Babylonian mathematical astronomy and the otherness of the rest of cuneiform knowledge. It also exported the nature/culture dichotomy to the cuneiform scribes and their corpus of knowledge. The faultiness of this particular dichotomy, not only within cuneiform sources but in general, has been mentioned already, discussed by Latour (1993) and shown to be problematic from a biological and environmental perspective too (Inglis and Bone 2006, 272). In historical terms, one of the clearest cases of the intrusion of culture into nature, and the impact that cultural goals of knowledge have on images of the world, can be seen in the variable ways that phenomena have been classified. To focus on the cuneiform sources, a look at some instances of cuneiform classification proves illuminating for its cultural diversity from other systems.

The key to the modern reconstruction of cuneiform Sumerian and Akkadian was found in the many lists of words and cuneiform signs compiled by scribes beginning, in the case of the unilingual (Sumerian) thematic word lists, at the end of the fourth millennium BCE with the very inception of writing itself. Bilingual (Sumerian and Akkadian) word lists and sign lists appeared over the course of the third and early second millennia, and the lexical texts continued to be copied and preserved until the end of the tradition of cuneiform writing as an important part of the training and the collected knowledge tradition of cuneiform scribes (Veldhuis 2014). The cuneiform lexical corpus, in organizing and arranging words and writings in relation to one another, reflects a certain project of classification and categorization. The word lists are organized around classifications of words for things, including nonmaterial things such as professions, as in the so-called lú-lists (Civil 1969), where lú is the Sumerian word we translate as "man, person, human being" (CAD s.v. *amīlu* meaning 1). By means of the operative classifications of words, different categories of things are evident.

The lexical texts provide spellings (in syllabograms and logograms) of words for things relevant to the needs of administration and bureaucracy, among other things, grouping the writings of these words according to a variety of functional criteria, as Veldhuis (2014) has shown. Beyond the functional criteria, it is also the case, notably, that words for things were grouped in accordance with sets of things with shared characteristics. What constituted a characteristic can be addressed as a matter of function or use, even as related to its material substances (words for things made of wood, tin, hide, or clay, and so on), but other elements of classification also considered morphological features of things (such as having wings, or a tail, four legs, being male or female). Significantly, the word lists do not exhibit classification by material substance as an answer to the question of essential material properties, that is "of being." What something did, or what someone did with something, was how the substance of which a thing was made took on classificatory importance. These groupings can be identified with what Goodman called relevant kinds (see Introduction, note 2) or cultural kinds, where meaningful distinctions are drawn between certain things and not others.

The series in question is titled ur_5-ra = *hubullu* after its first line (Landsberger 1957). Sumerian ur_5-ra was to be translated as Akkadian *hubullu* "obligation" or "debt (with interest)" (CAD H, s.v. *hubullu* A). This compilation, ur_5-ra or Ura, was organized into the categories of trees and objects made of wood, reeds and reed objects, clay vessels, hides and objects made of leather, metals and objects of metal, domestic animals, wild animals, cuts of meat, stones, plants, fish, birds, clothing, geographical names, and edibles. It is in the criteria by which such classes of things were identified that we get a glimpse of a way of ordering and structuring the world of the scribes, and therefore the words for the things so ordered and structured.

Words were arranged by classificatory pre- or postposed cuneiform signs, graphic determinatives of the category to which the word belonged. The plant list of Ura 17 (Landsberger, Reiner, and Civil 1970, 77–124, updated in Veldhuis 2014 and the Digital Corpus of Cuneiform Lexical Texts or DCCLT)[4] compiled the words spelled with preposed Ú, as Sumerian ú = Akkadian *šammu* "plant," as well as other such graphic indicators, or determinatives, used for the writing of plant names, such as SUM "garlic," NUMUN "seed," UKÚŠ designating kinds of "melon" or "gourd," and other indicators of groups relating to plants in a variety of ways.

[4] The DCCLT is found at http://oracc.org/dcclt.

Thus the plant list begins, with the Sumerian word in the left column and the Akkadian equivalent in the right:

ú *šammu* plant
ú.ebur *dīšu* spring grass
ú.li.a MIN ditto
ú.šár.ra MIN ditto
ú.gug$_4$ *umṣatum* (a kind of grass or weed)[5]
etc.

Whereas the plant list may appear to be unified around the idea of plants as a "natural kind," what constituted the category "plant," denoted by the determinative Ú, was conceptually different from the modern Linnaean classification of the various phyla of Plantae by their reproductive structures. Of course, how could it be otherwise, but this is clear evidence of the historicity of even the most basic of assumptions, such as how or by what criteria do we classify plants.

In the wider context of the lists, the descriptive classifications denoted by determinatives were based upon considerations of form, appearance, behavior, or utility. The taxonomic system indicated by the determinative Ú had nothing to do with an interest or understanding of plant reproduction as present-day plant classifications do but rather of outward form (leaves, stems, etc.) and habitat (field, marsh, etc.). This comes as no surprise, of course, but what I mean to emphasize here is that, as reflected in the case of Sumerian lexical lists, relevant kinds are a function of the purposes for which such sets of things are identified by those who need or are interested in them. Obviously, without the conception of nature at the disposal of the scribes, natural kinds – groupings defined as preexisting in nature – did not "exist" in their conceptual world or their conceptual vocabulary. Other kinds, however, did exist by furnishing that world with things, sets, and categories that point to other ontological orderings or criteria for such orderings of the world.

A different series of related but nonlexical texts were devoted to describing plants and stones, titled respectively "the plant, its appearance" (*šammu šikinšu*; Stadhouders 2011, 2012) and "the stone, its appearance" (*abnu šikinšu*; Schuster-Brandis 2008). These show that appearance, outward form, and usage were the criteria that grouped plants and stones as materials for pharmacological and magical healing purposes (Rochberg 2016a, 85–102 and 2019). Particularly in the *šikinšu* text for stones, the relevant characteristic of the words marked by the determinative NA$_4$ (*abnu* "stone") was that

[5] CAD, s.v. *umṣatu* B.

they were used to make beads, not only of stone but also of bone, shell, or metal, for healing purposes. The form and function of these "stones" comprised their groupings, not their intrinsic material substance or how they were formed "in nature" by geological processes, such as we distinguish by our classification of igneous, sedimentary, and metamorphic rocks. This can only be the case given that the order of nature was not something yet conceived of and so had no purpose for the cuneiform scribes' classification system. In spite of the obviousness of the fact that the ontological order of nature was a concept foreign to the Babylonian scribes, the cuneiform mode of classification of things resembling "natural kinds" (such as the various determinatives used for plants, animals, stones, etc.) in these as well as in the lexical texts can still sometimes be seen in contemporary literature as a prefiguring of, or rudimentary stage on the way to, a naturalistic physical classification system (Selz 2023, 105 and 114).

The classification system of the cuneiform lexicographical project fit the interests and purposes of the scribes who compiled the word lists, just as Linnaeus's *Systema Naturae* fit his and his contemporaries' naturalist aims. Cuneiform lexical lists organized words for things into categories relevant to scribal concerns (whether administrative or otherwise, such as medicinal plants for drugs), as well as having the practical function to translate Sumerian words into their Akkadian equivalents. Words for things that were organized together in relevant sets were construed in ways that sometimes but not always included the material constituent of a thing, but analogical considerations of what a thing resembled by virtue of certain features, such as having wings or being like a snake, also played a significant role in the development of the lists. As the lists had a taxonomic function by virtue of how words for things were organized, the question remains as to the ontological quotient of the lists.

The material constituent of the thing to which a word referred could be a criterion, such as in the classification of waterways under the word "river" (ÍD), but this group of words included natural waterways, such as rivers, as well as manmade waterways, such as canals. Similarly, things whose primary constituent was wood (GIŠ) could include natural things, such as various trees, or manmade things, such as beds, chariots, and doors. A different kind of classification, however, is reflected when the determinative, say "stone" (NA_4), indicates a bead made of various materials, not only of stone but also shell or metal. There the word "stone" (NA_4) is used as a determinative, but the things in the class marked "stone" (NA_4) are not all made of stone but belong to the class "stone" (NA_4) not by virtue of what we call their *nature*, but by virtue of their function, after having been made into beads.

The way graphic determinatives constructed sets of words for things did not require that membership in a class be based on necessary and sufficient features of the set members. As such, determinatives do not represent a monothetic type of classification. Where members of monothetic classifications are all identical as a result of their possessing both necessary and sufficient properties, those of polythetic classes belong to the group by other criteria of likeness (Rochberg 2016a, 99). Cuneiform determinatives utilized a polythetic classification system, as can be illustrated by the determinative "bird" (MUŠEN), which is appended to the names of birds, for example, $TI_8^{MUŠEN}$ = *alerû* "eagle," and $UGA^{MUŠEN}$ = *āribu* "raven." But MUŠEN, which means bird, is also used in the spelling of words for winged insects, as in [NIM].GEŠBUR(U)^{MUŠEN} = *nambūbtu* = *adammumu* "wasp." MUŠEN is preserved as well in the spelling of the names of constellations whose imagery was conceived of as birds, as in $^{MUL}TI_8^{MUŠEN}$ to spell the name of the constellation Aquila, the Eagle. The determinative MUL "star" also accompanied the writing of the name for Aquila, just as $^{MUL}UGA^{(MUŠEN)}$ is the writing for *Āribu*, the constellation Corvus, the Crow.

The determinative MUŠEN is thus a part of the spelling of the name of the constellation and has no purpose in classifying what the thing named "is" in a material sense, in this case, a constellation of stars. The feature of having wings, or flying, that is, what something "does" or resembles, is the function of the determinative MUŠEN. In the writing of the name of a constellation, the determinative only indirectly refers to a flying creature but is still considered necessary for the classification of an imaginary picture of a bird formed by stars.

Despite the fact that consideration for the material constituents of things is not the primary criterion of classification as indicated by the cuneiform determinatives, analysis of the classification of words and things by reference to determinatives does entail considering the relationship of classification to ontology in the cuneiform world and as such brings us back to the question of kinds. Kinds vary with context, from cultural or social kinds, or the relevant kinds of Goodman to the so-called natural kinds. Classificatory terms act as markers for an ontological system and how that system is structured, for example – remaining for the moment on the theme of birds – the (Linnaean) class of Aves.

According to the online Encyclopedia of Life, under "birds," "Aves is the Latin name for the birds – feathered, winged, bipedal, warm-blooded, egg-laying, vertebrate animals with evolutionary origins among the reptiles. … *In order to make classifications reflect evolutionary history* [my emphasis], they are now more usually regarded as falling inside the Reptilia."[6] Joseph Laporte noted that "the 'stem' of the birds, or Aves, is the species that gave rise to all of the known species of birds . . . probably some species of Archaeopteryx. . . . The relevant clade [meaning the species and all of its descendants] includes, in any possible world, all and only the organisms in and descended from species A [= Archaeopteryx]" (2004, 12, 66). He also pointed out that "cladistics systematists are now counting sparrows and ducks as less frightening models of the dinosaurs. Because birds evolved from the dinosaurs, they *are* dinosaurs" (Laporte 2004, 67). Ontologically speaking, for us, birds are classed according to their biological descent, not, as in the Babylonian system, for their winged appearance or their ability to fly, which allows insects and birds (and constellations in the shape of birds) to be classed with the same orthographic marker.

The framework for the classificatory work of cladistics is biological evolution and descent from a common ancestor. This framework also assumes that the classes of things it organizes exist as such, prefigured in nature, and are there to be discovered by biological science. The same applies to the plants, or categories of Plantae, for which their classifications depend upon an understanding of reproductive structures and therefore the evolutionary relationship between the phyla according to Linnaean taxonomy.

The example of the reclassification of birds as Reptilia points to the inherent weakness in a historiography of science whose aim is to decide when someone get nature's own structures wrong and when they get them right. For the history of science, the role played by *the purpose of classification*, evident in the case of the reclassification of birds as dinosaurs to make classifications reflect evolutionary history, is the critical piece of the historical puzzle. Behind all classifications are the reasons or purposes served by the making of the classifications themselves. As said before, it is not only functional but also ontological – the modern order of things depends upon naturalism and biological or geological evolution, the Sumero-Akkadian one does not and indeed could not.

The classification of things in the cuneiform world, as reflected in features of the writing system, it goes without saying, was not oriented toward a schema of nature such as we find in the Linnaean taxonomy or in geological classifications of rocks and minerals. In the kinds of things relevant to the cuneiform scribes for their compendia of words and

[6] http://eol.org/pages/695/overview.

signs, cuneiform culture *made* a world that differs from our material world of science with its goal to know nature and to understand biological or geological evolution. If, as Goodman said, "worlds differ in the relevant kinds they comprise" (1978, 10), then the cuneiform scribes' world, comprised by the relevant kinds found in lexical texts and indicated by cuneiform determinatives, differed from our own. The difference lies in the relationship between a world-version and the reality to which the version, with its particular classifications and kinds, refers. It is the relationship at the very center of the way of worldmaking.

In the following section, I explore the question of worldmaking through a key feature of modeling in cuneiform astronomical texts, and what it indicates about the objectives of these models.

NEXTNESS AS AN INSTRUMENT OF BABYLONIAN WORLDMAKING

The present section, by considering some of the modeling practices underpinning Babylonian astronomy, discusses the objectives of such modeling and how this has to do with worldmaking. It will be seen that the central objective is not representation of the physical motion of bodies in the heavens, but the recurrence of cyclical heavenly phenomena, and the ability to make predictions as a result. The focus in mathematical astronomical models on recurrence and repetition, which is the principle of nextness, is consistent with the aims of other forms of cuneiform scribal knowledge. This is not to say that nextness is an idea present in all the texts of *ṭupšarrūtu*, but its presence is a noteworthy feature of a good many such texts, especially those of the omen compendia.

Models aimed at representing astronomical phenomena operate within worldmaking frameworks. We can readily see this in Greek astronomical cinematic models, because a cinematic spherical model achieves a description of the motion of a planetary body within the enclosed space of the cosmos.[7] Babylonian astronomical models are not counterparts to this sort of model but utilized mathematical modeling to represent synodic phenomena[8] cycling through the zodiacal circle (counted in

[7] These Greek models were referred to as *hypotheses* (see Bowen 2020, 71–94) and as he points out, another difference between the Babylonian and the Greek models was that the Greeks found it important to justify and validate their models whereas the Babylonian astronomers did not.

[8] "Synodic" refers to the positional relationship between the sun and the moon, or the sun and a planet, from the Greek *syn* + *odos* ("the roads come together"). For the outer planets Jupiter and Saturn, which will illustrate this discussion, the synodic phenomena are first

degrees) in their various temporal periods (counted in lunar days or mean *tithis*, a unit the equivalent of $^{1}/_{30}$ of a mean lunar month).[9] The object of Babylonian astronomical models was the individual synodic phenomena of particular celestial bodies, or indeed quantities necessary for calculating the date or position in the zodiacal circle of some given celestial body in some given visible phenomenon.

Therefore, instead of accounting for the motion of the moon as a discrete body in space, Babylonian lunar models were designed to answer questions such as "When will the moon next be seen in its first visibility?" or "Where in the zodiac will the moon next be full?" The planetary models answered similar questions, such as "Where in the zodiac will Jupiter be when it is in its next first stationary point?" or "When will Jupiter be seen again in its first stationary point?" These questions more than suggest that the motivation in modeling celestial phenomena for Babylonian astronomers was not a concern with physical properties of motion but rather with a nonphysical property of what can be descriptively called nextness.

If the structure of the tables can also be taken to reflect the conception of the behavior of the synodic phenomena, nextness can be understood not only as a formal feature of the tables but as an ontological property of the astronomical *bodies* that recur and thus exhibit periodicity; here, *body* refers not to the physical substance of the moon or planet, but only its appearance in a specific synodic moment. As a nonphysical property, nextness is instantiated in the recurrent planetary and lunar phenomena, such as an outer planet in its first station, or an inner planet in its first morning visibility. The nextness of such bodies is, as already said, not physical, but has to do with order. For celestial bodies, it is the order we recognize as periodicity.

The common characteristic of the Babylonian models is calculation of the next phenomenon in periodic sequence, be it the length of the day or the date of a synodic phenomenon of the moon or planets, generated as consecutive entries in columns. Not only were dates and zodiacal positions of consecutive synodic moments entered in individual columns of the

visibility, first station, second station, and last visibility, followed by conjunction when the planet is not visible due to its being in the same direction in the sky as the sun. Between the first and second stations is the retrograde arc, when the planet appears to move "backward" against the fixed-star background or against the direction of the zodiacal signs.

[9] Neugebauer (1947, 146, note 6) credited the discovery of the use of the "mean lunar day" in the Babylonian ephemerides to A. Pannekoek (1916) and again independently to B. L. van der Waerden (1941).

tables, but also in some cases strictly mathematical quantities (such as differences between the values in different columns) comprised a column enabling one to proceed from a value in one column down to the next entry in the adjacent column.

Before proceeding with a discussion of the modeling of phenomena in the tabular mathematical astronomical texts, it should not go without mention that another form of so-called nonmathematical astronomy also stemmed from the premise of nextness. This is now termed goal year methods, which made rough prediction of a number of planetary and lunar phenomena of the kind recorded in the Diaries, both synodic and sidereal (Hunger ADART vol. 6; Brack-Bernsen 1999; Brack-Bernsen and Hunger 2008; Gray and Steele 2008; Gray and Steele 2009). The phenomena of interest to the goal year texts are identified in the colophons as "the first day, appearances, passings [i.e., planetary distances to certain ecliptical fixed stars], and eclipses, which have been established for year x," meaning the dates of these phenomena are found by using their characteristic periods to project the next such phenomenon by citing the phenomenon in question from the Diaries one period before the target, or goal year. In what follows, however, I focus instead on the tabular methods and how they shed light on the conceptual framework for Babylonian representation of celestial phenomena through modeling.

It is common to describe the character of the Babylonian tables, and thus of Babylonian astronomy more generally, as predictive (e.g., McMullin's P-science, 1984a, mentioned in the Introduction). This is one way of viewing what is certainly a chief characteristic of Babylonian mathematical astronomy and its aims, but it seems that we can go further. Nextness does not preclude or substitute for mathematical prediction, but for the present exploration it is worth noting that, as an idea distinct from prediction, nextness is found throughout the corpora of *ṭupšarrūtu* that organized knowledge in lists. Characteristic of the majority of such lists was a vertical consecutive relationship between one entry and the one that followed it, at least within delineated sections, whether in lexical lists or lists of omens, although horizontal relationships from one column, or entry, to the one following it horizontally also obtained (in Old Babylonian extispicy, Winitzer 2017, 151–231; in lexicography, Geller 2022).

The mathematical prediction of the dates and positions of synodic phenomena was the outcome of the interest in nextness but does not in itself capture the formal structure of these tables or explain the Babylonian conceptualization of astronomical periodicity. The nuance of nextness is that it describes a formal characteristic, not only of the mathematical astronomical tables, but also of other corpora from the tradition of *ṭupšarrūtu*. And in the case of astronomy, it seems also to relate to a deeper conceptual aspect, which was an understanding of planetary and lunar recurring synodic phenomena. Models for the recurrence of those phenomena were utilized for predictive purpose, but nextness operates on a different register from prediction, having to do with a particular way of worldmaking.

Why should nextness be taken as an instrument of worldmaking? Were Babylonian modeling practices simply alternatives to Greek astronomy such that we can assume a shared conception of the phenomena and a shared conception of the world? Do Babylonian and Greek astronomical models merely stand as different representations of the same phenomena? These questions recall the worldviews-versus-worlds, or the one-ontology -versus-multiple-ontologies debate discussed in Chapters 4 and 5. They also recall a statement by Kuhn to the effect that we are in for a shock when we replace our own conceptual spectacles for those of another culture, even, or perhaps especially, when natural phenomena are in question. In his words, "the shock generated by substituting their conceptual spectacles for our own is the one I ascribed, however inadequately, to their living in a different world" (Kuhn [1991] 2000b, 221).

If we peer through the conceptual spectacles on the world that differ with history and culture, scientific model-making can be seen to reflect diverse world-versions with varying ontological frameworks. Put another way, scientific models attach at many points to the world-versions to which their communities of inquirers are committed. The geocentric spherical models of Greek astronomy were firmly attached to a certain philosophical and cosmological framework, just as the mathematical ones of Babylonian astronomy came out of millennia of working with lists and tables, as well as of producing mathematical schemes to describe periodic events such as the length of the day over the course of a year, or the duration of the visibility of the moon over the course of a month, both attested long before the period of the Late Babylonian mathematical astronomical texts. Because from one corpus to another the criteria determining the inner logic of the cuneiform lists will differ, the remainder of this section explicates nextness in the context of astronomy so as to further clarify this argument.

Evidence for the quantitative modeling of astronomical phenomena is extant from the entire chronological span of Babylonian astronomy. The evidence without exception takes the form of arithmetical constructs to denote, describe, and predict cyclical astronomical phenomena. The cycles in question varied depending on the phenomenon of interest to the scribes and their desire to know when and in what position (celestial longitude in degrees of zodiacal signs only in the late Babylonian astronomy) the next phenomenon in the periodic cycle of the body in question would occur. Commonly, models were constructed to generate cyclical phenomena such as the length of the day, the duration of the moon's visibility at night, a first visibility of the moon, an eclipse, or a synodic planetary phenomenon. The chronological range of this practice extends from the Old Babylonian scheme for the variable length of the day to the Neo-Assyrian scheme for the visibility of the moon in *Enūma Anu Enlil* Tablet 14 to the Late Babylonian ephemerides of the Seleucid period for planetary and lunar periodic phenomena.

The mathematical structures applied to the calculation of such phenomena were constructed on the basis of sufficient empirical inputs for the purpose (de Jong 2019), but the dates and positions tabulated were subject to varying theoretical constraints, such as the calculation schemes of Systems A and B, the mathematical models of late Babylonian astronomy (see Glossary of Astronomical Terminology Cited). By means of such modeling, the quantities generated by such mathematical schemes had a denotational relationship to some phenomenon of the world,[10] such as a length of daylight, a lunar eclipse, or a synodic appearance of a planet, as mentioned before. A variety of mathematical constructs, represented by the different "systems" (A, A′, A″, B, B′, B″, and K, as described in Ossendrijver 2012), obtained the desired results.

Babylonian astronomical modeling was based on parameters for relevant periods, the units in which those parameters were expressible, as well as available coordinates, such as degrees of zodiacal longitude (counted as 1–30 degrees of a zodiacal sign). Systems A and B used the step or zigzag functions (see Glossary of Astronomical Terminology Cited), respectively, to account for the difference between a position/longitude and the next in periodic sequence. In System A, this difference was modeled as the progress of a phenomenon around the zodiacal circle subdivided into zones of longitudinal progress, the simplest version consisting of two such zones,

[10] Goodman (1968, 5) saw denotation as essential to representation but not necessarily dependent on resemblance. The elements of a model, from this standpoint, need not replicate or simulate the object of the model in a physical way, but if there is the fundamental relationship of reference, or denotation, between model and world, then that model represents the phenomena in the observer's empirical *and* theoretical world.

or arcs (α) of the ecliptic, one fast and one slow. In this way, System A modeled the synodic arc – the characteristic change in celestial longitude between successive synodic events ($\Delta\lambda$) of the same kind – with a step function of longitude (Aaboe 2001, 44). The object of System A was not to describe the motion of a celestial body in a physical sense but to use what we describe as "fast" and "slow" arcs of the ecliptic (in a two-zone model) as a mathematical way to deal with the problem of anomaly manifest in real as opposed to mean synodic arcs. In making the computation of consecutive phenomena dependent upon the position of the body in the zodiacal circle, System A's calculation scheme was tied to the zodiacal positions so that the change in longitude was a direct function of longitude ($\Delta\lambda = f(\lambda)$).

System B, on the other hand, used a period tied to the serial number (n) of a synodic phenomenon in some sequence of consecutive synodic phenomena in an ephemeris table (Aaboe 1964, 225). But since both systems used the same basic period relations between Π phenomena (the "number period")[11] and Z years, their mean synodic arcs are the same, for example, Jupiter's mean period of 33;8,44,48° underlies both Systems A and B in spite of the different ways of mathematically accounting for the intervals (δ) or "steps" between occurrences of the phenomenon (Neugebauer 1975, 420–31). The values of the synodic arcs ($\Delta\lambda$), tabulated in a System B ephemeris, increase and decrease between two extrema and thus form a linear zigzag function ($\Delta\lambda = f(n)$), where the independent variable is the serial number of the consecutive events to which those synodic arcs belong.

Despite the fact that the two systems modeled synodic phenomena with different mathematical schemes, the schemes theorized how one appearance of a phenomenon followed another in the same way, that is, by means of the intervals (δ) or "steps" between phenomena and the definition of true synodic arcs as $\Delta\lambda = Z\delta$. The differences in arithmetical structure between Systems A and B were the result of how the steps δ were distributed, that is, either as a stretchwise constant on multiple arcs (k) of the ecliptic (System A) or as a continuous arithmetical progression increasing and decreasing between extrema by a constant difference (System B; Neugebauer 1975, 421). The critical feature of a Babylonian mathematical model is not the resemblance between the model and the phenomenon it denoted in the way a Greek cinematic model simulated

[11] That is, the number of phenomena that occurred within some integer number of cycles around the zodiacal circle.

the motion of a planet around the geocentric celestial sphere. And yet the interest in nextness and the derivation of periods and relations between periods to get a fix on the positions and dates of recurrent phenomena is derived from direct engagement with the world conceived within the Babylonian version.

The components of the Babylonian modeling of planetary phenomena were, as mentioned before, the number of phenomena (Π) and the number of rotations of the zodiacal circle (Z) required for an integer number of phenomena to occur. The period relation was expressed in terms of an integer number of phenomena (Π) and its corresponding period counted in an integer number of rotations of the zodiacal circle, or years (Z). One can also derive the period P from the relation Π/Z, which results in a value representing the number of synodic phenomena that will occur over a single rotation of the 360° of the zodiac. This makes for a troublesome physical concept for us, because in no case is the value of Π/Z an integer (see the table of values of Π/Z in Neugebauer 1975, 423). Π/Z for Jupiter is the irreducible fraction 391/36, which counts the number of times the phenomenon repeats (391 times) in the course of the number of sidereal rotations[12] of the zodiacal circle to bring it back to its initial position (36 times around). But the period, defined as one rotation of the zodiacal circle (P = Π/Z) results in the value 10;51,40 phenomena, according to System A. We then see that not quite eleven synodic phenomena of the planet Jupiter, say a first station, occur in exactly one rotation of the 360° of the zodiac.

To relate the period, as represented by the number of phenomena, to the time in years for these to occur, Neugebauer explained:

> Experience must also have shown that such a period [the number period Π] corresponds to an integer number of sidereal rotations of the planet and therefore also to an integer number of years since all planetary phenomena of the same kind occur in the same relation to the sun, a fact easily established by observing the relative motion of the planet and of the sun against the background of the fixed stars. (1975, 420)

Neugebauer considered the nature of these planetary period relations, as follows, to be empirical (1975, 420):

Π + Z years or Z sidereal rotations of Saturn or Jupiter
2 Π + Z years or Π + Z sidereal rotations of Mars
Π + Z years or sidereal rotations of Venus
Z years or sidereal rotations of Mercury

[12] The term "sidereal rotation" is used because the Babylonian zodiac was sidereally fixed, meaning fixed by the stars without regard for the precession of the equinoxes.

Therefore, the modeling of planetary synodic appearances comes from relating the number of synodic phenomena, that is, appearances of a given synodic phenomenon, and the number of sidereal rotations of the zodiacal circle by the body in question needed to obtain an integer number of appearances. Π + Z years is the total time required for the planet Jupiter in some given synodic appearance to return to the same zodiacal longitude of its initial position in that same appearance. As Jupiter's number period (Π) is 391, in the course of performing 391 synodic appearances of some given phenomenon, Jupiter – as some given synodic appearance – makes 36 sidereal rotations and the total number of years for Jupiter's synodic cycle is Π + Z = 427 years. In 427 years, therefore, Jupiter will perform 391 synodic appearances of a given phenomenon (selecting for a first visibility, a first station, and so on) and will have completed 36 rotations of the zodiacal circle to return to its initial position.

Babylonian astronomical modeling is effected by denoting a visible phenomenon but without a physical, material, or mechanical-causal structure. Moreover, Babylonian astronomy's goal was not to offer a simulacrum of the behavior of celestial phenomena as motion through space. It developed tools for generating (predicting) the sequence of synodic appearances, that is, the phenomenon y_{n+1} from a value y preceding it in the table. Thus, Babylonian astronomical modeling was denotative and representational without attempting to describe or simulate reality mechanically or account for it materially. The empirical basis for such modeling, however, maintained the connection between the models and the world of the scribes' experience, understanding, and imagination.

The foregoing description of Babylonian astronomical modeling was meant to highlight its character as against other forms of astronomical modeling, such as in Greek astronomy. This is evident from early (seventh century) arithmetical models of the change in the length of the day (in the Babylonian Astrolabes, Horowitz 2014) or the duration of lunar visibility (George and al-Rawi 1991) to the later tables. All have in common the use of mathematical modeling to derive the next appearance of a phenomenon of interest. As already stated, nextness is not the same as prediction, although prediction can be an outcome of it. Looking to celestial omens, nextness was also of key interest in spite of a lack of quantitative periodicity, although many ominous celestial phenomena were certainly periodic, such as eclipses. Even the nonperiodic phenomena among ominous phenomena (colors of the planets and other events) were of interest whenever they would next appear. On that basis, Babylonian mathematical astronomy was consistent with other parts of cuneiform knowledge.

By viewing the Babylonian astronomical calculation systems in light of the wider tradition of knowledge to which they belonged, it follows that underpinning the tables were cultural conceptions with ontological implications, if only we can ferret them out. We get no purchase on physical properties or material substance from the Babylonian mathematical models. However, if a property such as nextness serves a similar function to matter or motion as a feature of celestial phenomena in the Babylonian view, why should we not grant that property ontological significance within the world as framed by the scribes? Nextness does not necessarily preclude an interest in the material substance of physical entities, but the cuneiform contexts in which nextness is a dominant property of things give no indication that physical differences between things was important.

To further argue for models employing nextness as an instrument of worldmaking, a look at the formulation of what a model is in relation to a world by the iconic cognitive and computer scientist, Marvin Minsky, cofounder of the Artificial Intelligence (AI) laboratory at MIT, will illustrate (Minsky 1965).[13] My use of Minsky's formulation considers how his way of defining a model applies to worldmaking. In this short paper, Minsky discussed the relation between mental and physical events and how models mediate between the mind and the world. He defined a model in the following terms, that "to an observer B, an object A* is a model of an object A to the extent that B can use A* to answer questions that interest him about A" (1965, 45). Minsky saw the intention of the knower, or observer B, as crucial to the use of A* (the model) in relation to A (something in the world). The purpose of a model is then crucial to how successful it is at answering to the actual phenomena it models. He added (1965, 45; my emphasis), "if A *is* the world ... A* is a good model of A, in B's view, to the extent that A*'s answers agree with A's, on the whole, over those questions important to B." Therefore, Minsky defined models as inherently ternary, that is, comprised of the triad of model A* to a phenomenon A for an observer B.

In Minsky's schematic of modeling, observer B's role is paramount, being first the source of the identification or selection of phenomena to be modeled, and second the source of the questions the model is designed to answer. But B's questions are mediated by her culture, that is, whatever it is that makes B interested in A and prompts her to formulate the questions she does about A. Minsky also allowed for the possibility that A is not

[13] I thank Willard McCarty for drawing my attention to this reference.

merely a phenomenon or set of phenomena, but the world itself, in which case A* becomes a representation of the world (A) inhabited and investigated by B.

In light of worldmaking then, Minsky's definition can be reformulated as follows: If instead of saying A may not merely be a phenomenon or set of phenomena denoted by model A*, but the world *itself*, we can redefine the world A not as the fixed and found ready-made world but rather the world as conceived and made by the community of inquirers in question. The denotata of model A*, then, are the phenomena comprising the world as determined in and by the historical, cultural, or ideational context of those for whom model A* is successful.[14] The model A* becomes a representation of A *as it is made* by B. Inherent to the process of modeling is, therefore, the triadic relationship of observer, model, and object/or world.

Models of phenomena epitomize scientific knowledge, at once denoting, describing, representing, or in some cases explaining how or what the world is thought to be in accordance with a governing epistemology and its ontological commitment. The Babylonian mathematical astronomical models were designed to capture the nextness of cyclical phenomena, enabling prediction. And because the periods and period relations were good, the Babylonian models testify to the rigor and "rightness" of the Babylonian world-version.

By means of the empirical underpinnings of these mathematical models, a fundamental relationship of reference was effected between the models and the Babylonian world. An armillary sphere effects a model of the heavens in a different way from that of Babylonian Systems A and B by being a physical simulacrum of a spherical cosmos. Earth-centered spheres were models of the Ptolemaic conception of the cosmos, sun-centered devices were Copernican, but both presupposed the finite celestial sphere. Babylonian Systems A and B modeled planetary and lunar phenomena without necessitating the same kind of physical relationship to the cosmos, but the conceptual relation between the observer and the appearances of heavenly bodies accounted for the success of those models, no matter if we judge them to be accurate or inaccurate. In the

[14] This schematic should not be read as rigid or static. Traditions of knowledge transcend cultural identities and languages to persist across vast chronological spans, as the history of science in the Hellenistic world certainly exemplifies. However, it is worth considering to what extent ideas are modified in the transmission of science to accommodate or repurpose old models within different ways of worldmaking.

case of the armillary sphere as a physical simulacrum of a spherical cosmos, whether earth-centered or sun-centered, the world so modeled is as much a construct as that of the Babylonian models. The major and important difference is that the former denoted the cosmos as understood by a dominant Greek and then early modern European astronomical tradition, while the latter denoted a set of periodic phenomena without reference to a cosmological frame but operating within a Babylonian conception (version) of the world. In each case, however, the conceptions that produced these variant astronomical models were instruments of worldmaking.

After looking in detail at the Babylonian style of model-making from the evidence of the mathematical astronomical systems of calculation we designate as Systems A and B, and considering the character and aims of that form of astronomy, it seems to me to be an injustice to the history of scientific modeling to classify those that do not operate within the familiar boundary lines of naturalism, materialism, or mechanism as merely representing worldviews and not an actual world, which depends on and cannot be wholly separated from the version attached to it. We have long ago accepted the quotient of theory that influences observation and empirical data (theory-ladenness). It follows that scientific representation, classification, and ontological frameworks are laden with a quotient of cultural and historical diversity (version-ladenness).

We cannot, therefore, understand how the ancients knew, or sought to know, the phenomena of their world without also asking what features they attributed to that world and how they identified, defined, and organized the phenomena they selected as objects of their inquiry. In the context of astronomy specifically, it is clear that if cultural differences are fully taken into account, Greek cinematic spherical model-making and Babylonian mathematical model-making are not reducible to a single system. Nor was Babylonian astronomy a rudimentary Greek astronomy or Greek astronomy an undeveloped early modern European astronomy (cf. Berlin 1982, 108).

THOMAS S. KUHN ON WORLDS

To conclude this chapter, and bring the discussion back to questions of the tension between cultural diversity and human (or natural) universality, I turn once again to Thomas Kuhn's reflections on worlds, world-versions, and ontological difference. These are subjects Kuhn explored in two essays: first "Possible Worlds in History of Science" ([1989] 2000, 9–32) and then "The Natural and the Human Sciences" ([1991] 2000b). I review what is relevant to our present concerns from his "possible worlds" essay first.[15]

> To possess a lexicon, a structured vocabulary, is to have access to the varied set of worlds which that lexicon can be used to describe. Different lexicons – those of different cultures or different historical periods, for example – give access to different sets of possible worlds, largely but never entirely overlapping. . . . Anything which can be said in one language can, with imagination and effort, be *understood* by a speaker of another. (Kuhn [1989] 2000, 61)

Kuhn ([1989] 2000, 59–60, 63–65) was interested in what he called possible-world semantics, where true statements in one world may not be true in another. This is very like Goodman's discussion of contradictory "truths" in conflicting world-versions (Goodman 1978, 110).[16] The difference is that Goodman was not interested in possible worlds but rather actual worlds answering to conflicting versions (Goodman 1978, 2, 20, 95, 120). Between two of Kuhn's possible worlds, untranslatability resulted in incommensurability, more specifically, as "applied to a pair of theories in the same historical line, the term meant that there was no common language into which both could be fully translated" (Kuhn [1989] 2000, 60).

In a twist on Quinean indeterminacy of translation, in which there is no unique meaning for a word or even a whole sentence because of a lack of any absolute objective criterion or standard for establishing the correct translation of a given speaker's meaning, Kuhn took the opposing position, saying, "instead of there being an infinite number of translations compatible with all normal dispositions to speech behavior, there are

[15] Hitherto, citations will be from the Conant and Haugland 2000 republication.

[16] What it is not like, is David Lewis's idea of possible worlds ([1986] 2001). Apart from the differences between Goodman's plurality of ways of worldmaking and Kuhn's possible worlds, it should be clear that neither one meant the same thing as what Lewis meant by the plurality of possible worlds. Lewis's worlds were a function of modal realism, viz., when we imagine how the world could or might have been. When Lewis said, "absolutely every way that a world could be is a way that some world is" (1986] 2001, 86), he populated a modal "space" of many conceivable (and inconceivable) ways things might be. But, clarifying modality versus actuality for his many worlds, he stated, "ours is the actual world; the rest are not actual ([1986] 2001, 92). In reference to the "actual world," Lewis was a monist. As I read him, the actual/real world is only one certain way. Plurality he reserved for the many ways "things might have been different" ([1986] 2001, 1). He addressed neither Goodman nor Kuhn in his book, and indeed there were no touch points, at least none made explicit by Lewis. In the context of pluriworldism and history, I do not find Lewis's modal realism to warrant further discussion.

often none at all" (Kuhn [1989] 2000, 61). Accordingly, one cannot translate from one scientific lexicon to another as long as their respective terms refer to different "worlds" in which different laws apply. But, significantly for the historian, "the impossibility of translation does not, of course, bar users of one lexicon from learning the other. And having done so, they can join the two together, enriching their initial lexicon by adding to it sets of terms from the one they just have acquired" (Kuhn [1989] 2000, 74).

The second of Kuhn's essays relates even more directly to the uses of diversity and the irreducible nature of different constructional systems over the course of the history of science. In view of the different taxonomies for kinds of celestial objects employed by the ancient Greeks, he stated, "the heavens of the Greeks were irreducibly different from ours" (Kuhn [1991] 2000b, 220). Because the heavenly universe seems unchanging and absolute, it has been a favorite example of how nature is independent of any human interpretation of it, and the stars are always "the same." But Kuhn's statement estranged what was hitherto assumed to be known and familiar, namely the heavens, by affirming their difference in different "worlds." With this statement, Kuhn raised the question for the history of science of the meaning of "the same." This very question then became a critical element of Goodman's *Ways of Worldmaking*. He did not refer to Kuhn directly but said that to say of two things, for example, the heavens of the Babylonians and those of the Greeks, that "both are versions of the same facts," is not to be taken "to imply that there are independent facts of which both are versions" (Goodman 1978, 93). This line of reasoning, as previously discussed (Chapter 4), derives from Goodman's irrealism, in which he saw the interdependence of fact (world) and convention (world-version) and the impossibility of either existing independently of the other.

Kuhn referred to a conceptual rather than an imaginary "brute data" vocabulary as what makes the crucial difference between worlds. With clear implications for our ability to resolve incommensurable cultural or historical lexicons and ideas, precisely what Lloyd would later call mutual intelligibility, he added, "that does not mean that one cannot, with sufficient patience and effort, discover the categories of another culture or of an earlier stage of one's own" (Kuhn [1991] 2000b, 220). And, significantly, "it does indicate that discovery is required and that hermeneutic interpretation – whether by the anthropologist or the historian – is how such discovering is done" (Kuhn [1991] 2000b, 220). Far from setting two world-versions apart in unbreachable realms of thought, Kuhn saw analysis and interpretation as laying a path to enter the conceptual vocabulary of the other. Directly relevant to our concerns with ethnocentrism in the historiography of science, he said:

> The violence and misrepresentation consequent on describing their [the Greeks'] heavens in the conceptual vocabulary required to describe our own is an example of what I then called incommensurability. And the shock generated by substituting their conceptual spectacles for our own is the one I ascribed, however inadequately, to their living in a different world. Where the social world of another culture is at issue, we have learned, *against our own deep-seated ethnocentric resistance*, to take shock for granted. We can, and in my view must, learn to do the same for their natural worlds. (Kuhn [1991] 2000b, 221; my emphasis)

Incommensurability, therefore, was an indication of the plurality of worlds. It did not pose an insurmountable barrier to understanding, but access had to be gained through "discovery." Some of Kuhn's ideas are clearly in the background of those of Geertz, Rorty, and Lloyd, specifically in the critique of the dualistic idea of the cultural superstructure as ephemeral overlay upon a unitary fixed reality.

Kuhn, Goodman, Geertz, Rorty, and Lloyd all came to see "the uses of diversity" from different yet interrelated perspectives and considered the implications of such diversity for ontological pluralism. Goodman's project of worldmaking stemmed from philosophical inquiry and from questions developed in large measure from his interest in Cassirer's epistemology ("In just what sense are there many worlds?" and "How is worldmaking related to knowing?" Goodman 1978, 1) and the philosophy of symbolic forms ("What role do symbols place in the making?" Goodman 1978, 1), as opposed to something like the philosophy of modal realism and its many possible worlds (Lewis [1986] 2001). Kuhn's ideas stemmed directly from historical inquiry into the sciences, Lloyd's from both the history of ancient sciences and historical anthropology. Geertz, of enormous influence in cultural (symbolic) anthropology, was, however, challenged by Rorty and then too by the anthropological "ontologists" for his representationalism (Paleček and Risjord 2012, 7–8) and treatment of symbol systems.

The relevance even today of Goodman's worldmaking to historical sciences, in spite of his insistence that science was not the exclusive way of making worlds, lies in the fact that it supplies a powerful way out of ethnocentric historiography. The value of the anthropology of science in the context of *ṭupšarrūtu* is as a method of inquiring into the world answering to the cuneiform sources. To employ this method of inquiry is one of the uses of diversity, that is, so that we can attend to the modes of description and conceptualization reflected in cuneiform scientific texts, and from that point of departure make their world intelligible to us in ours. In the next chapter, the conceptual lexicon of the cuneiform texts in which we find world descriptions provides an opening for apprehending how the world answering to those descriptions was conceived.

6

Worldmaking in Cuneiform Culture

He who through vast immensity can pierce,
See worlds on worlds compose one universe,
Observe how system into system runs,
What other planets circle other suns,
What vary'd being peoples every star,
May tell why Heaven has made us as we are.
But of this frame, the bearing and the ties,
The strong connections, nice dependencies,
Gradations just, has thy pervading soul
Look'd thro? Or can a part contain the whole?
– *Alexander Pope,* An Essay on Man, *Epistle i*

FROM COSMOLOGY TO WORLDMAKING

For the following discussion of ways the world and its parts were expressed in Akkadian, some clarification is necessary for the different senses in which I take the terms "cosmology" and "worldmaking." In some contexts, surely, they have overlapping connotations, but in others not at all. Cosmology has come to refer to the systematic study, resulting account, and representation of the physical universe. In historical terms, cosmologies differ and even depict conflicting worlds (the earth-centered and the sun-centered). Worldmaking encompasses historically divergent cosmologies but goes one step further to challenge the assumption that the differences are merely products of worldviews, as has been noted throughout this book.

As discussed in the previous chapter, the strain between diversity and universalism has not gone away, and positions on universalism still vary within anthropology and the social sciences more broadly.

Indeed, Pina-Cabral's (2014a, 2014b) universalist hopes for anthropology were coupled with a strong commitment to ontological monism, reserving plurality for worldviews, not worlds. The issues raised in his two papers signaled to me that it *is* still useful to consider the range of meaning in historical conceptions of "world" in premodern non-Western (meaning non-Greek) cultures. In particular, the terminology and description of the world and its parts in cuneiform texts is the aim of this chapter.

Cosmology, by virtue of its etymology from Latin *cosmologia*, means "discourse on the cosmos" (Greek *kosmos* "universe," Latin *mundus* "world") whereas, as a gerund construction, "worldmaking" has a processual-substantive connotation that can refer not only to "the cosmos" as the object of a way of worldmaking, but also, or instead, some other conception of a lived environment in whatever way that may have been construed. That is to say, worldmaking does not always signify cosmology as an end goal but works on various levels of human dealings with the phenomena. Cuneiform sources lend themselves more to an investigation of worldmaking than they do of cosmology, for reasons I hope to make clear in the following discussion.

Cuneiform worldmaking is suggested by certain themes running through many texts from the corpora of *ṭupšarrūtu* and from literary texts that draw upon its learned traditions. These texts reflect themes such as complementarity and counterparts, correspondence, measurement, and proportion that function more to describe the relation between world parts and less to describe the world as a body with a certain shape and discrete boundaries.

The sources in which we find descriptions of the world and its parts range from literary texts about gods and their domiciles, whether in heaven or in the underworld, to descriptions of the levels of the created world and of the systematic arrangement of the celestial bodies and the gods they represent. Because of the extraordinary chronological range of the relevant sources, it is not possible to reconstruct a single unified or even consistent version of the cuneiform world construction. Wayne Horowitz ([1998] 2017) assembled the sources having to do with the topography of the imagined celestial world but judiciously steered clear of unifying them into a consistent "cosmology," or world picture.

The cuneiform texts, be they narrative and mythological or astrological and astronomical, did not share the same objectives in depicting the world as are found in Greek philosophical texts. Orderliness and justice are ideas that obliquely connect Babylonian (as well as Sumero-Akkadian and Assyro-Babylonian) formulations to the Platonic definition of cosmos, which figure in *Gorgias* 508a. There, Socrates says: "Wise men say that heaven and earth and gods and men are bound together by communion and friendship, orderliness, temperance, and justice and it is for that reason they call this Whole a Cosmos" (Allen 1984, 297). What cosmic order and justice, not to mention a concept of "this Whole," amount to in cuneiform traditions, however, is altogether different from that in the Greek tradition. Within Sumero-Akkadian, Assyro-Babylonian, and Babylonian sources, the idea of cosmic concord, which came to be a central point in Greek cosmologies, was refracted through the role of the gods, seen as active in the world. While the cuneiform scribes never converted their ideas about order and concord into material or cinematic properties of matter, or of spherical geometry, they expressed proportion and measurement of the world's parts in a variety of ways, enough for us to identify these as recurring themes in cuneiform worldmaking.

Cuneiform worldmaking, reconstructed from Sumerian and Akkadian mythology, hymns, celestial divination, and astronomical and related texts, can be discussed in terms of the creation, structure, and workings of various parts of the world. Conceptions of the world as a whole comprised of parts, of heaven and earth and their relationship to one another, and ideas about the nature of the heavenly bodies emerge from diverse texts. A way of worldmaking according to cuneiform culture does not emerge by asking whether the celestial bodies were thought to move in a vacuum or whether earth was fixed or not. These are questions that stem from a particular form of Western cosmological inquiry, wherein the overriding interest was in the physical nature of earth and the heavens, their respective material constituents, and the motion characteristic of their distinct kinds of matter. Instead, Sumero-Akkadian poets, mythographers, and scholars, touched on questions about the world, its parts, and their relational structure in the form of hymns of divine praise, poetic compositions, and their commentaries, as well as in lists of celestial omens and their scholia. The poem *Enūma Eliš* "When above," which relates the creation of a new world by the god Marduk, is one such source. As Lambert stated:

> Other than *Enūma Eliš*, there is no systematic treatment of cosmology in Sumero-Babylonian literature. . . . But this does not mean that *Enūma Eliš* presents all that is known of Babylonian cosmology. On the contrary, the Epic uses only a selection of the wealth of available material . . . parallels to Marduk's work have to be collected from allusions and incidental comments.
>
> (Lambert 2013, 169)

The Classical world produced more than one cosmological text and more than one cosmology (Pythagorean, Milesian, Platonic, atomist, Aristotelian, Stoic, Ptolemaic). Atomists, such as Democritus and Leucippus, were said to have had the idea of innumerable worlds (Kirk et al. 1983, 416–19, with texts 563 and 565). Cuneiform texts do not entertain the idea of multiple worlds in quite the same way but attest to multiple heavens (*šamû* [*plurale tantum*]) and earths (*erṣetu*) in one seventh-century Assyrian tradition (extant in two sources):

> The upper heaven is of *luludānītu*-stone; it is (the heaven) of Anu. He (Bēl "lord") settled the 300 Igigi therein.
> The middle heaven is of *saggilmud*-stone; it is (the heaven) of the Igigi-gods. Bēl took up residence there on a sublime dais. On a lapis lazuli dais he made it (the middle heaven) shine from within with *būṣu*-glass and crystal.
> The lower heaven is of jasper, and the constellations of the gods are drawn upon it.
> On the [grou]nd of the upper earth, he allowed the spirits of mankind to rest (after death).
> [On the grou]nd of the middle earth, he settled Ea, his father . . .
> [On the ground of the] lower [earth], he confined the 600 Anunnaki-gods.
> (KAR 307: 30–37, and AO 8196 iv 20–22, *Archiv für Orientforschung* 19 pl. 33, Horowitz 1998, 3–4, translation modified)

The three heavens depict levels in the heavens made of different stones to describe their colors, or in the case of the middle heaven, its radiance: the multicolored *luludānītu*-stone of the highest heaven, not visible to humankind, the middle heaven of a different stone (of unknown color) known to have been suitable for making beads against "everything evil" (*mimma lemnu*) (CAD, s.v. *saggilmud*, usage b), and the visible blue jasper color of the lower heaven, where humankind sees the constellations "drawn" on its surface.

When it comes to the question of world structure (Lambert 1975, 2013; Horowitz [1998] 2017; Geller 1999–2000; Katz 2003), the key text is the Babylonian poem *Enūma Eliš*. The text was recovered in 1849, when Austen Henry Layard found the clay tablets containing the seven-tablet poem in the ruins of Nineveh. In 1876, George Smith brought out an English edition, and in 1890, Peter Jensen's *Die Kosmologie der Babylonier* set the text within the wider scope of cosmology. At the start of the twentieth century, the British Museum published L. W. King's cuneiform copies of the text in volume 13 of *Cuneiform Texts from Babylonian Tablets in the British Museum* (1901). King's edition and translation, titled *The Seven Tablets of Creation, or the Babylonian and Assyrian Legends concerning the Creation of the World and of Mankind*, appeared in rapid succession in 1902. The authoritative edition appeared in Lambert 2013. This poem is perhaps the finest example of cuneiform literature and scribal scholarship, in terms of not only its high poetic style but also and notably its scholarly mastery of cuneiform orthographies and hermeneutics. It belongs at the center of any discussion of Babylonian world structures according to first-millennium scribes.

When we look at the narrative of creation from *Enūma Eliš*, especially in Tablet IV, we see that the story is principally a moral one, while also explaining the structural components of its world. Marduk, the hero, slays the watery goddess Tiamat, whose body will become the heavens. In killing Tiamat, Marduk vanquishes evil, treachery, and improper divine rule. Tiamat was guilty for appointing her own spouse, Qingu, to "the rank of Anushıp" (the highest divine station) though he had no right to that office. She is further accused of giving Qingu the Tablet of Destinies,[1] though he had no right to hold it. She is charged with stirring up trouble and perpetuating evil against the gods. In the battle, Marduk is aided by an incantation for undoing evil (Akkadian *tû*, Sumerian TU_6) as he approached his foe:

> He set his face toward the raging Tiamat. In his lips he held a spell (*tû*) and in his hand he held a plant to counteract (Tiamat's evil).
>
> (*Enūma Eliš* IV 60–62)

Tiamat, as well, when she found she was up against the redoubtable Marduk, first turned into a raving maniac: "like one insane she lost her mind" (*Enūma Eliš* IV 88, *mahhûtiš ītemi ušanni ṭēnša*) and then repeatedly recited an incantation to protect herself (*Enūma Eliš* IV 91). The moral victory was complete when Marduk rounded up Tiamat's divine helpers, smashed their weapons, threw them in a net, and confined them to prison (*Enūma Eliš* IV 111–14). The culminating triumph was when the hero Marduk wrested the Tablet of Destinies away from Qingu, whose possession of it had been an infraction against divine propriety (*Enūma Eliš* IV 121).

[1] The mythological Tablet of Destinies is defined this way by Andrew George: "The theological exposition of the function and nature of the Tablet of Destinies ... makes use of the traditional terminology associated with the tablet. It is ... the means by which supreme power is exercised: the power invested in the rightful keeper of the Tablet of Destinies is that of chief of the destiny-decreeing gods ... which amounts in principle to kingship of the gods. The poet of the Anzu Epic makes this clear when he describes Anzu's usurpation of supreme kingship by theft of the tablet, and the concomitant upsetting of the cosmic status quo" (George 1986, 138).

When after the Babylonian *Chaoskampf* the creation of the world was finally mentioned in *Enūma Eliš* IV 135–36, the god Marduk rested and surveyed the corpse of Tiamat, which had been rendered a lifeless marsh (*sarkuppu/serkuppu*) or lump (*kūbu*).[2] Having then fashioned the heavens from half of her tightly stretched body, and keeping her waters from escaping by installing watchmen, Marduk proceeded to make heaven a counterpart of the subterranean watery Apsû, precisely measured to its size. Finally, Marduk rectified the order of divine propriety by setting up shrines where the high gods were to have their residences (*Enūma Eliš* IV 141–46). The major parts of the world that were created in order to house these deities, the "great gods," and set the world aright were the heavens (home to Anu); earth (as Marduk's new residence); the Apsû (home to Ea), also called Ešgalla; and Ešarra (literally "House of the All") for Enlil. *Enūma Eliš* provides an authoritative picture of the creation of the world, but variations in world structural imagery are also attested (Rochberg 2020), as seen in the Assyrian text of the multiple heavens and earths.

Because *Enūma Eliš* is a key source for themes of worldmaking, the following discussion focuses on its contribution to world description rather than its hermeneutic acrobatics that showcase the highest level of scribal techniques of writing and reading (van de Mieroop 2016, 4–9). In his 2016 study, van de Mieroop discussed *Enūma Eliš* Tablet VII (edition in Lambert 2013, 122–34), where the fifty names of Marduk are intoned (Tablet VI 121, "Come, let us call the fifty names!"; Lambert 2013, 117:121). The fifty names consist of a summation of everything the creator god established for the purpose of world order and rectitude, and they are a tour de force of Sumero-Akkadian philology and layered meaning. Van de Mieroop's analysis of the poem captures its moral and practical system, as well as the lofty hermeneutics of the written symbol, the writing of cuneiform signs, words, and names, or, as he put it, "the universe as a written text" (van de Mieroop 2016, 11).

In his analysis, van de Mieroop posits a dyad of world and text, consistent with the structuralist approach in other aspects of his book. Certainly, worlds need words (and/or images) to describe (and depict) them, much as Saussure's signifier and signified stand in mutual and dyadic relation to one another. Evidence of the importation of thought about world phenomena into the writing system was discussed in the

[2] The word *sarkuppu/serkuppu* can also be read as ^{UZU}*kūbu* "lump" (of flesh), which usually refers to a stillborn or premature fetus, or a monstrosity of some kind. A *double entendre* was no doubt intended.

previous chapter in relation to the classifying determinatives used in the cuneiform writing system. Other aspects of this orthographic phenomenon, characteristic of the cuneiform writing system, were brought out by van de Mieroop as well (2016, 8–9). The primarily philological nature of the scribes' endeavor, he hypothesized, meant that the text determined reality, in his words, "writing preceded reality" (2016, 224). The following offers a different angle of approach to an interpretation, consistent with the ternary relationship of scribe, text, and world.

Because the present discussion focuses on the description of world parts and their interrelationships, it proceeds from a different construal of the relationship between the written text and the world, which is the object of its reference and signification. Instead of a dyad consisting of world as fixed external reality on the one hand, and text as its signifier on the other, I propose adding to the meaning-making process the phenomena of the world as a functional part of the observer's construction of a world in relation to them. Therefore, instead of two, I posit three elements involved in the process of signification, namely, the observer (in this case, the scribe), the world (the phenomena of cultural interest that are the objects of observation, perception, conceptualization), and the text (or visual representation, or the model, as discussed in Chapter 5). The idea of mutually constructive worlds and world-versions is a useful navigational instrument for conversion of a fixed dyadic relationship of text to reality on the one hand to the ternary relationship just described.

The ternary, or triadic, relationship in question parallels that of model construction discussed before (Chapter 5), where three elements contribute equally to the construction and success of modeling, namely, the observer (i.e., the observer's interests and questions), the phenomena (or world as experienced and understood by the observer), and the model (the representation of the phenomena or world as seen or understood by the observer). Moving from the dyad of world and text to a triad of world, observer (together with her time, place, language, and culture), and text (or model), the world that stands in relation to a text is but one of three interdependent parts of a relational process by which the observer makes sense of her lived environment. This is the framework within which I propose to discuss some themes of worldmaking in cuneiform texts in the following section.

THEMES OF WORLD DESCRIPTION

As Goodman said in the introduction to *Ways of Worldmaking*, "we are confined to ways of describing whatever is described" (1978, 3). How the first-millennium Babylonian scribes understood the structure and composition of their world is largely described in mythological narratives and hymns to particular gods (see Horowitz 1998; Katz 2003; Lambert 2013). In Goodman's sense, the scribes' descriptions of the world and its parts construct a world-version that is rendered right for its frame of reference.

Mythological narratives and god lists of cuneiform culture have been approached as representative of the rudimentary beginnings of the history of cosmology (Lambert 1975). The world-version or versions attested in the stories of gods and in the poem of world creation (*Enūma Eliš*) was once criticized as unprogressive and leading nowhere for a "scientific cosmology," as in the following statement by Lambert:

> In many ways it is a disappointment that the civilisation which produced so much in the collection of data and in the abstract sciences did not develop its cosmological ideas during the 3,000 years of its existence, and the reason for this can be sought. Two main factors were at work. One was the old, prehistoric mythological thinking, which, despite the fictional element in its products, was properly based on sound observation of the universe. This could have provided the basis for a scientific cosmology if conditions had been right. The other was the personification of natural forces into anthropomorphic gods and goddesses, and this led away from the realities of nature into theological fantasy. Unfortunately the growth of civilisation resulted in a diminution of the first of these two factors and an increase in the second.
>
> (Lambert 1975, 49)

Lambert found that cosmological ideas were stunted for thousands of years because of "old, prehistoric mythological thinking" and "personification of natural forces." This statement, a relic of scientism, separated science (as the "realities of nature") from religion (as "theological fantasy") into a stale dichotomy no one subscribes to any longer.

Those "mythological narratives" are still centrally important to a study of worldmaking rather than cosmology, which, following Goodman, looks to the system of description rather than what is described for a different analytical point of view. Accordingly, my objective is to consider how the ways of describing used by ancient cuneiform scholars evoke the world of their making. In looking for the ways the scribes described their world, or parts of their world, it is possible to discern a complex of interrelated themes having to do with relationships between interrelated parts, such as regions in the heavens, on earth, in the netherworld, and in "the abyss" (*Apsû*) (Lambert 1999–2000). An identification of persistent themes in sources spanning many centuries and across distant geographical areas becomes a productive way to reach some of the ideas inherent in these sources, even when they are not explained or made explicit.

In all cognizance of the variation in descriptive representations, with especially wide variation diachronically, a thematic analysis is pursued here, based upon Akkadian meronyms (names for parts) for the purpose of understanding how Akkadian set parts in relation to a whole (meronomy), especially when the whole is likely construed as "the world." This analysis raises the question of how, or in what way, out of the many terms for "all of x (parts)," the notion of a whole was constructed. The concept of a whole that is of interest here, of course, is that of "the world," or the "totality" of everything. Primary among the themes that play a role in the relations between parts of the world are complementarity, counterparts, and correspondence. Because these indicate various ways of relating named parts (meronyms) to each other, how a whole might be conceived out of parts, and how a unity is made of a multiplicity of such interrelated parts, they are indicative of a way of worldmaking.

The focus on meronymy (names of parts) and meronomy (the part–whole relationships) points up another contrast between the cuneiform evidence and more familiar approaches to understanding the cosmos in the Greek philosophical world. Even in diverse cosmological schemes in ancient Greece, the part–whole relation is clear in the notion of *kosmos* that underpinned the ancient Greek way of worldmaking. A case in point is Aristotle's definition, or rather three definitions (*De Caelo* 1.9.278b9–21),[3] of *ouranos* from the work *On the Heavens*, which concludes in the following passage: "Since we use *ouranos* in three senses, the entire world surrounded by the outermost revolution has to be formed from all physical, perceptible body because there is neither anything outside of the outmost revolution nor can there be" (*De Caelo* 1.9.278b21). Another locus for the idea of a unified cosmos is found in Stoicism. As Brooke Holmes pointed out, "the conspicuous fact of cosmic sympathy compels us in turn, the Stoics insist, to accept the reality of less obvious things, such as the unity of the body of the cosmos" (2019, 245). She quotes Sextus Empiricus (from *Against the Professors* 9.79) as saying that from such observations as the phases of the moon coordinating with phenomena on earth, "it is obvious that the cosmos is a unified body" (Holmes 2019, 245, note 17). Holmes further points to the Stoics' "pious commitment to the unity, completeness, and perfection of the world together with

[3] These are (1) the substance of the outermost revolution of the universe of the world and the seat of divinity; (2) the body containing the moon, sun, and some of the stars; (3) the body enclosed by the outermost revolution, i.e., the entire world, that is, the universe, *ouranos* (see Leggatt 1995, 88–89).

a tarrying with its parts and its particulars" (2019, 240).[4] In marked contradistinction, the cuneiform descriptions of the relations among world parts did not focus on unity, completeness, or perfection, but rather on correspondences, counterparts, and complementarities. As such the cuneiform way of worldmaking stands apart from the entire complex of natural philosophical thought.

For conceptions of the world expressed in cuneiform texts, we have no definitions as such but rather elements of descriptions from a variety of sources, such as literary texts, divine epithets, names of temples, and royal inscriptions. The world parts named in those genres also appear in the corpus of *ṭupšarrūtu*, in omen texts and commented passages in scholastic Akkadian explanatory texts. One of the more interesting, and puzzling, aspects of the evidence overall is that rarely is reference made to the world as an integral whole or a place with explicit outer boundaries. However, whereas we find words for parts (meronyms) of *the one whole* (we would say "world" or "universe"), we do not find a word for that one unified whole, the holonym meaning "world" but instead a word for "totality, all, or entirety (of parts)."

In each case, the Akkadian words meaning "all," "totality," or "entirety" are constructed with another noun in the genitive for a sense of "all of" some set of things included within it or counted as parts of it. For example, a word meaning "all" or "totality" (*kiššatu*) is used precisely in the sense of "all of x," or "the totality of x." In these constructions *kiššatu* is found together with some other noun in the genitive plural, as in "all (of the) lands" (*kiššat mātāte*), "the entire sky" (*kiššat šamê*), "all mankind" (*kiššat nisī*), or "all the gods" (*kiššat ilī*). Alone, the term *kiššatu* connoted "the entire inhabited world," as in the well-known royal title "king of the universe" (*šar kiššati*), where *kiššatu* has the sense of all known inhabited places. Divine epithets for the gods Marduk and Nabû use the phrase "all heaven and earth" (*kiššat šamê u erṣeti*) as a designation of worldwide power, but the meaning is conveyed by the construction of "all of" with the merism "heaven and earth." Another example is the phrase used in "When Above" (*Enūma Eliš* IV 14) when the gods gave kingship to Marduk. There they delegated him to rule over "the entirety of the whole of everything" (*kiššat kal gimrēti*).

An interesting aspect of the term *kiššatu* is that it can be written with the logogram ŠÁR (*šá-ár* ŠÁR = *kiš-ša-tum*; Idu 2.70; CAD, s.v. *kiššatu*

[4] Other Greek philosophical traditions, notably the atomists and especially the Epicureans, entertained the idea of innumerable worlds (*kosmoi*), as discussed by Warren (2004).

lexical section), which, as the word for the number 3,600, is further used to mean an indefinite large number, from Sumerian šar (ePSD, s.v. "3600"). The Akkadian loanword *šār* means "countless," and so by extension "all," as in the phrase *šār bēri*, literally "3,600 double-hours" used to mean "a countless number of x" (CAD, s.v. *šār*, meaning 1b). The extended use of *šār*, however, is limited to the expression *adi šāri* to convey a sense of "for all time" or "everywhere," mostly in bilingual literary contexts. These expressions convey a sense of temporal as well as spatial "totality," as in the passage that refers to the radiance (*melammû*) of kingship that overwhelms "everywhere" (*adi šāri* in CAD, s.v. *šār*, meaning 2).

In a note on the origin of the Greek term "saros" (σάρος) for the eclipse relation 223 synodic months = 242 anomalistic months, Neugebauer (1969, 141) pointed to its derivation from the Sumerian numeral šar for 3,600, and he suggested a meaning: "universe or the like." As a very large number word, Sumerian šar/Akkadian *šār* became "saros" (σάρος) for the period for prediction of eclipses every approximately eighteen years (6585;20 days as a standard value), but that would have come entirely from the Greek perspective, as there is no astronomical significance to Sumerian šar or Akkadian *šār*. On the other hand, the earliest graphic form of Sumerian šar was a circle (see Figure 6.1), which may have had this shape as a representation of the idea of an encompassing "all" or "totality."

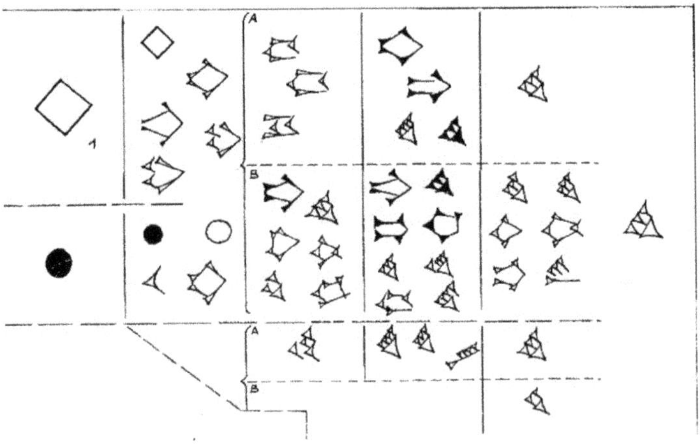

FIGURE 6.1 Evolution of the cuneiform sign ŠÁR (Labat 1988, 180–81). Photo courtesy of Editions Guethner.

Another word, *kullatu*, similarly, means "all," or "totality," and is frequently constructed with a genitive of parts that can be counted, for example, "all the gods" (*kullat ilī*), "all peoples" (*kullat nišī*), or "all cities" (*kullat māhāzi*). In extended meaning, there is "the whole of all wisdom" (*kullat nagbi nēmeqi*), and perhaps most general in semantic force is the construction with the idiom "everything, anything," literally "whatever its name" (*mimma šumšu*), to mean "all of everything" (*kullat mimma šumšu*). In this case, "whatever its name" (*mimma šumšu*) is an abstract way of designating "all parts" and so is similar to the previous examples where *kullat* + gods, lands, or peoples is the way to express "all of x."

The substantive *kalu* similarly means "whole, entirety, all" and is sometimes constructed following a noun referring to things that can be counted, as in "signs, all of them" (*ittātim kalašina*) or "prediction (literally, "verdicts") from birds, all (kinds) of them" (*purussû iṣṣūrū kalašunu*). The extended meaning of the word *napharu* "all, whole, totality" stems from its basic meaning "sum, total (of things counted or added)," which occurs in the context of a total tally of items in a list, or the sums of numbers (CAD, s.v. meaning 1). It is sometimes found in the absolute, for example, "the totality of people, all of them" ([*te*]*nēšetu ša naphari kališunu*), or to describe rulership over (*ana malikūti*) everything, or one who is appointed to be in charge (*pāqid*) of everything (CAD, s.v. meaning 2, usage a), or, as mentioned before, rulership over "the entirety of the whole of everything" (*Enūma Eliš* IV 14). More often it is found in construction with *kullatu*, as in "the totality of everything" (*kullat naphari*), also preceding a genitive such as "all of Assyria" (*naphar māt Aššur*), "all of Chaldea" (*naphar māt Kaldi*), or "the totality of gods" (*naphar ilī*), and "all the black-headed (people) as many as exist" (*naphar ṣalmat qaqqadi mala bašû*).

Finally, a poetic word *nagbu* means "totality" or "all" in the same way as its synonyms *kullatu*, *napharu*, and *kiššatu*.[5] Just as in the case of its synonyms, *nagbu* (CAD, s.v. *nagbu* B) refers to all of something, constructed with what is being totaled in the genitive, such as "all the Aramaeans" (*nagab Ahlamê*), "all his enemies" (*nagab zā'erīšu*), "all the swamps" (*nagab berate*), and "all living beings" (*nagab napišti*). Already mentioned is also "the whole of all wisdom" (*kullat nagbi nēmeqi*).

[5] It is included in the lexical synonym, or thematic, word list Erimhuš (5.44) and is given as a synonym of *napharu* in the commentary to the "Akkadian Theodicy" (Lambert 1960, 74, commentary to line 57).

Somewhat different is the use of the term "regions" (*kibrātu*; CAD, s.v.), often constructed in the expression "the four quarters" (*kibrāt erbetti*), which then represented "the entire inhabited world." In one of its usages, the phrase "the four quarters" (*kibrāt erbetti*) has a political rather than a cosmological connotation, used in royal titulary and in inscriptions describing either hostilities in or domination over "the four quarters." Common is the expression that the king "has no rival among the princes of the four quarters" (*ina malkē ša kibrāt erbetta šāninšu la išû*). That the four quarters represent the entire world reflects a royal ideology in which divine appointment to kingship is in fact to rule over the entire world, as in the claim to be the one "whom Assur appointed to rule the entire world" (*ša Aššur … ana mu'irrūt kibrāt arba'i šumšu … išquru*; CAD, s.v. *kibrātu*, usage a, 1'b').

Also relevant are *gimru* and *gimrētu*, from the verb *gamāru*, whose basic meaning ranges from "to finish" to "to encompass" and "to complete."[6] The derived forms *gimru* and *gimrētu* present a somewhat different sense of totality compared with the previously discussed terminology. One such passage is from the *Akītu* ritual: "You (Marduk) survey everything with your eyes, you explore further all omens/oracles/decrees/extispicies (*têrtu*) through the *omina* you give" (*ina īnēku tabarri gimrētu* [*ina*] HAR.BAD.MEŠ-*ka tahâṭu* HAR.BAD.MEŠ; RAcc. 130:20). Another reference to *gimrētu* (*gimrātu*) is in a lexical synonym list entry as *gimratum* = MIN (= [*puhru*]) (CT 18 21 Rm. 354:7). The noun *puhru* (under *puhru* A in CAD) stems from *pahāru*, meaning to gather things together. In another passage for *puhru*, "Anu created all (meaning the collection of all) the heavens, Anu created (the collection of) all the earths" (*Anu puhur samê Anu puhur erṣeti … ibtani* Köcher, BAM 538 ii 52) equally connotes the plurality of things making up "the whole." Like *kullatu*, *puhru* "all" or "totality" takes its meaning from being constructed with the genitive of its component parts.

Finally, *kippatu* "circle" has the meaning "totality" when constructed with world parts such as lands (*kippat mātāte* "all lands") or the four quarters (*kippat erbette* "all four quarters of the earth," CAD, s.v. meaning 3c). Its more common usage is in construction with the merism "heaven and earth" to mean "all the world," usually in divine epithets for gods who hold all the world (*tāmih šamê u erṣeti*; CAD, s.v. meaning 3) or hold the bond of the whole world (*markas šamê u erṣeti*).

[6] Also "to bring to an end," as in "annihilate." See CAD, s.v. *gamāru*.

The foregoing examples underscore the difference between the concept of a totality expressed by "all *xs*," and that which functions within a lexical part–whole hierarchy, or meronomy. In the fields of semantics and pragmatics, according to D. A. Cruse, a meronomy is a "type of branching hierarchy" of the "part–whole type" (Cruse 1986, 157). Cruse gives as the prototype for meronomies the division into parts of the human body, where the whole in that case is the body occupying the apex of the hierarchy, and underneath are parts such as "head," "arm," and so on. These, then, have their own parts that branch underneath, such as "nose," "hand," and so forth. In the cuneiform texts, instead of finding a hierarchy of the whole to its parts, we more often find the idea of a sum of parts.

The Akkadian expressions for part–whole relations do not convey the same vertical hierarchies as Cruse analyzed in the semantics of the English language (1986, 157–80). Of the Akkadian terms for "all of x (parts)" discussed here, none offers a true holonym for the whole of what the parts are comprised. Given Cruse's assertion that "there are no meronomies of unnamed wholes" (1986, 171), it is difficult to know how to analyze the various world parts (heaven[s], earth[s], netherworld, Apsû) in the absence of a name (holonym) for the whole of the world itself. I turn now to the themes that relate the parts to each other and, perhaps then indirectly, indicate a notion of the whole.

COMPLEMENTARITY AND COUNTERPARTS

As outlined in the previous section, passages containing the many Akkadian terms for "all" (*kullatu*, *kiššatu*, *kalu*, *napharu*, *nagbu*, *gimru*, *gimrētu*, *puhru*, and *kippatu*) do not focus their descriptions on the structure or material substance or overall shape of the world as a whole, but rather as the sum of its many parts or constituent things. In contradistinction to a meronomic branching hierarchy, we find references to parts of the world in various nonhierarchical binary relations to one other, most basic of which are the regions of heaven(s) and earth(s). The idea that certain parts of the world, such as "above" and "below," complement one another, forming by virtue of their relation one whole, is an essential ingredient of cuneiform worldmaking. The use of merisms, composed of terms for contrasting or complementary parts, for example, "heaven and earth," is common in Sumerian and Akkadian (cf. in the Hebrew Bible, discussed in Noegel 2017, 123–37). Parts that complement one another, perfecting and making a whole from those parts, are often at the same time viewed as counterparts of one another.

The complementarity of two principal parts seems to be the oldest and most basic conception of "world" in the cuneiform tradition. The idea is preserved in the terms AN.TA.MEŠ (Akkadian *elâtu*) "the upper region(s)" and KI.TA.MEŠ (Akkadian *šaplâtu*) "the lower region(s)," and the idea of upper and lower halves of the world reflects an original mythology of creation by the separation of AN "above" from KI "below" (Lambert 2013, 169–70). The complementary parts, heaven and earth, were a pair in Sumerian as an and ki "heaven and earth" (as well as the Sumerian an-šár and ki-šár or "all heaven" and "all earth"). In Akkadian, the merism "heaven and earth (or underworld)" (*šamû u erṣetu*) remains the most basic structural duality for the world as a whole.

The merism "above and below" is, moreover, basic to the way the world is made through celestial divination, namely, that signs above are equivalent to signs below. Celestial and terrestrial signs are counterparts, as expressed in a singular text from the seventh century known as the Diviner's Manual, which puts it as follows: "The signs on earth as well as of the sky bear (lit. "wear," *našû*) omens[7] for us; heaven and earth alike bring us omens; they are not separate from one another; heaven and earth are interconnected (*itḫuzu*). A sign that is unfavorable in heaven is unfavorable on earth; that which is unfavorable on earth is unfavorable in heaven" (Oppenheim 1974, 200: 38–42, translation modified). This describes a world wherein the counterparts of fortune and misfortune are manifested equivalently in the two complementary realms of the world, heaven above and earth below.

The complementary status of heaven and earth is complicated by the fact that the "below" of this equation (the lexical equation KI = *erṣetu*) denotes both earth and netherworld, depending on context. Sources confirm the identification of the region called *erṣetu* as the place where netherworld deities reside; thus "earth" means "netherworld" in some contexts (Horowitz 1998, 273–74), while in others *erṣetu* means the dwelling place of the living, the seat of the god Enlil. The literary text *Atra-Hasīs* says "I cannot [set my feet on] the earth of Enlil" (Lambert et al. 1969, 91:48).

The complementarity of heaven and earth/netherworld (*erṣetu*) is also found in a motif repeated in temple hymns and royal inscriptions concerning the building of ziggurats and temples, whose tops were so high as to

[7] This line uses two different words for "sign" or "omen," *ittu* and *ṣaddu*, the latter signifying especially signs produced by deities, including the celestial deities manifested as planets (moon, Jupiter). See CAD, s.v. *ṣaddu*.

"rival heaven" and whose foundations were so deep as to reach the netherworld (*erṣetu*). As an example, "he (Sargon) established its (the temple Eanna's) foundation in the depth of the netherworld like a mountain" (*temmenšu ina irat kigalla ušaršid šadū'aiš* YOS 1 38 i 40). It was also a way of expressing great magnitude, as in the description of a mountain encountered by Sargon II (721–705 BCE) on campaign, "whose summit above leans against the heavens, and whose base, below, is firmly rooted in the netherworld" (*ša eliš rēšāša šamāmi endāma šaplānu šuršūša šuršudu qereb aralli* TCL 3 p.6:19). The motif was applied to mythological mountains in the Gilgamesh Epic that reached the base of heaven and extended down to the netherworld (George 1999a, 71, ix 38–41).

The heights and depths of a mountain or a temple had a variant in which earth/netherworld is replaced by the depths of the abyss itself. The metaphor is traceable to the early third millennium BCE in an archaic hymn (Biggs 1971, 201): "Great, true temple, reaching the sky, temple, great crown, reaching the sky, temple, rainbow, reaching the sky, temple, whose platform(?) is suspended from the midst of the sky, whose foundation fills the Apsû." The Apsû was the place of the watery abyss – not the salt sea, rather the sweet fresh waters of the springs (*nagbu*[8]), as we know from Sumerian and Akkadian mythology, divine epithets, and hymnic literature. The unceasing, seemingly eternal, nature of the freshwater springs and source of the rivers no doubt accounts for the inclusion of the Apsû as one of the regions of the world. Thus in the Hymn to Šamaš, about the man who does good deeds, it is said, "[His] descendants will go on forever like the eternal spring waters" (*kīma mê nagbi dārî zēra*[*šu*] *dā* [*ri*] Lambert 1960, 132:121).

The innermost part of the freshwater abyss was beyond the reach of human perception. No one had plumbed the depths of the Apsû nor seen its interior (Horowitz 1998, 317). The depth of the Apsû was an unknown region, as described in a broken line (from the portion of the clay tablet that had broken away) that compares something (in the broken beginning of the line) to the unknowable inner regions of the Apsû, saying "that, like the inside of the distant Apsû, not even a god can find (it)."[9] On the other

[8] This *nagbu* is a different lexeme from its homophonous *nagbu* B "totality," discussed at p. 179. For *nagbu* "spring" or "source," see CAD, s.v. *nagbu* A.

[9] See CAD, s.v. (*w*)*atû* lexical section for the passage "sù.ud.du.gin$_x$ dingir na.me.nu.mu. pàd.da.e.ne : ... *ša kīma qereb apsî rūqi ilu mamman la uttû*," translated there as "who, like the innermost of distant Apsû no god can discern."

hand, the brilliance of deities, such as Ninurta, is praised as illuminating the depths of the abyss. Although a part of the world largely inaccessible to human perception, the fresh waters of rivers and springs bubbling up from underground were places where the Apsû reached earth's surface.

Apsû's watery depths were associated with the creation and dwelling place of the god Enki (Sumerian)/Ea (Akkadian), so closely associated with him that Enki/Ea's son, Marduk, was known as "firstborn son of the Apsû." Because of Enki/Ea's association with wisdom, magic, and incantations, the Apsû was the fount of wisdom and source of the secret knowledge of incantations. The Temple of Enki in the oldest Sumerian city of Eridu was called the É-Abzu "House of the Abyss." In *Enūma Eliš*, Marduk's temple Esagil in Babylon was said to be the counterpart (*mihirtu*) of Apsû (*Enūma Eliš* VI 62).[10] *Enūma Eliš* (IV 142) places Ešarra, the dwelling place of Enlil, as a counterpart (*tamšīlu*) of Ešgalla, the "great shrine" or Apsû. And the ziggurat foundation, Etemenanki (its name "House, Foundation Platform of Above and Below/Heaven and Underworld" is itself testimony to the idea of the complementarity of above and below) is said to be a copy (GABA.RI) of Ešarra in the compilation TIN.TIR IV: 2 (George 1992, 58–59) and also in an inscription of Nabopolassar (also written GABA.RI; VAB 4 62 iii 19, cited in Lambert 2013, 200). *Enūma Eliš* further accounts for the location of the city of Babylon and the site of Marduk's own temple Esagil there, as a new world center (George 1999b, 67–86). Marduk's residence in the Temple Esagil effectively placed Babylon at the center of the vertical world structured with heaven and Apsû as the extreme limits. Marduk's "house," the Esagil, is said to be the equivalent, or counterpart, to the great abyss, the Apsû, where Ea dwelled: "They raised the peak of Esagil, a replica of the Apsû" (*Enūma Eliš* VI 62) using the term *mihirtu* "equivalent, counterpart."

The importance of the idea of measured counterparts, equivalents, is also clear in *Enūma Eliš* IV 141–46: "He (Marduk) crossed the heavens, surveyed the sky, and he adjusted them as the equivalent of the Apsû, Nudimmud's abode. Bēl measured the shape of the Apsû and set up Ešarra, a replica of Ešgalla. In Ešgalla, Ešarra which he had built, and the heavens, he settled in their shrines (the gods) Anu, Enlil, and Ea" (Lambert 2013, 95). A further indication that the idea of counterparts played a thematic role in a cuneiform world description is the use of the term *maṭṭalātu* "image/counterpart" from a verb (*naṭālu*) meaning "to look at, face, or point toward." The word

[10] Esagil is also said to be the GABA.RI *ap-se-(e)* "counterpart" (lit. "copy") of the Apsû in VAB 7 300 10, cited in Lambert 2013, 200.

maṭṭalātu is relatively rare, occurring only in first-millennium scholarly or literary contexts. The clearest usage is no doubt the one found as the incipit of Tablet 16 of the liver omen series (*Barûtu* "the art of inspection"): "If the liver is an image/counterpart of heaven" (*šumma amūtu maṭṭalāt šamê* CT 20 1:31). Another attestation of *maṭṭalātu* occurs in the account of the rebuilding of the temple Esagil by Esarhaddon. The rebuilding of Marduk's temple sanctuary signified Esarhaddon's political and ideological reconciliation with Babylonia. In his royal inscription detailing this act (see p. 196), reference is made to the Babylonian *Enūma Eliš* in describing the temple Esagil as the counterpart of the Apsû (*maṭṭalāt apsî*) and the equivalent, or likeness (*tamšīlu*), of Ešarra.[11]

In the context of world regions, the theme of the counterpart was applied to temples, mountains, trees, and the liver, connecting the divine, or heavenly, world with the human. When the temple of Enlil, the Ešarra, is set up as the replica or counterpart (*mihirtu*) of the Field Constellation, or the Square of Pegasus, the celestial quadrant marked by the stars is established as the model in heaven for the earthly shape and foundation for the Ešarra temple in Babylon. The best indication that the meaning of *mithurtu* is opposition (Goetze 1959, 125), in the sense of complementary counterpart, is in the writing of the Sumerian word that is translated into Akkadian as *mithurtu*, namely HA.MUN, something meaning "harmony (of opposites)" and is spelled with the Sumerogram NAGA.NAGA, where the second NAGA is written upside down:

CORRESPONDENCE

Continuing the nonhierarchical nature of themes attested in many world descriptions from *ṭupšarrūtu* is that of correspondence, which forms a different relationship from that of complementarity in pairing ominous signs with their consequents, the protases and apodoses of omen texts. Every omen instantiates the relation of correspondence, so this is perhaps the best attested of all the worldmaking themes in cuneiform knowledge.

[11] The reading could also be *mithurtu*, which has the meaning "opposition," as in *lišān mithurti* "contrasting tongues or conflicting opinions," in which case the word might be read *mithartu* "square." See the discussion section under *mithurtu* in CAD.

Typical examples of correspondence can be given where deities are set in corresponding relationships with phenomena, for example with the four winds: south (corresponds to) Ea, the father of the gods; east (corresponds to) Enlil, the lord of all; north (corresponds to) Ninlil, lord of phantoms; west (corresponds to) Anu, father of heaven (George 1992, 152–53). Trees are also correlated with body parts of gods, as in this Neo-Assyrian ritual text:

> Honey is the pus of the kidnapped god.
> Tamarisk is his topknot,
> Cypress is [his] trunk,
> Juniper is [his] thighs,
> Cedar is his knees,
> Medlar is [his] ankle bones. (Livingstone 1989, 38:9–17)

The parts of the liver correspond to specific deities, and the months to zodiacal signs, as in this example from a late text from Uruk:

> The Presence (corresponds to) Enlil, the month Nisan, [the Hired Man (Aries)]. . . . The Path (corresponds to) Shamash, the month Ajaru, The Bull of Heaven (Taurus). . . . The Pleasing Word (corresponds to) Nusku, the month Simanu, True Shepherd of Anu (Orion). . . . The Strength (corresponds to) Ninurta, the month Du'ūzu, the Crab (Cancer) . . . and so on.
> (SpTU 4.159, cited in Koch-Westenholz 2000, 24–25)

Correspondence between zodiacal signs and parts of the human body can be copiously documented in late medical astrology. Correspondences given in texts mentioning the microzodiac relate the twelfth parts of a zodiacal sign (*dodecatemoria*, Akkadian *zittu* "portion") to temples, stones, and woods (Heeßel 2005). Correspondences are attested in virtually all systems of cuneiform knowledge, in lexicography, omens, astrology, and medicine. Through correspondence, objects of knowledge of interest to the scribes were brought into relation with the world around, be it the heavens (*Enūma Anu Enlil*), the behavior of animals (*Šumma ālu*), plants (medical texts, late astrological medicine), or stones (medical and ritual texts, late astrological medicine). What is significant about these correspondences is the way in which the relationships were made, such as by semantic association, conceptual analogy, or orthographic relationships between corresponding parts.

A dramatic exemplar of correspondence, and also a clear instantiation of the way of worldmaking that joined human beings to the heavens, is the use by two Neo-Assyrian monarchs (Sargon II and Esarhaddon) of so-called astroglyphs for the writing of their names and titles (Figure 6.2).

FIGURE 6.2 Stone Prism of Esarhaddon (WA 91027) with glyphs for the name and titles of King Esarhaddon of Assyria, seventh century BCE. Previously published in Finkel and Reade 1996; Scurlock 1997; Roaf and Zgoll 2001; Morenz 2003. Photo courtesy of the Trustees of the British Museum

These illustrated inscriptions contain imagery, symbols that can be "read" as a glyphic script wholly separate from cuneiform. Esarhaddon makes the following statement:

I carved on them constellations, the image of my name (*lumāši tamšīl šiṭir šumjya ēsiq ṣēruššun*; Leichty 2011, 201: 11–12)

Not every glyph in the Esarhaddon texts can be understood as having an astral correspondence. Nonetheless, the use of symbols corresponding to constellations with which to represent the words in the royal name is found on the Stone Prism of Esarhaddon, as well as on the prisms with the circular arrangement of the same, or similar, glyphs (Roaf and Zgoll 2001, 266). The difference between the two is that on the Stone Prism the Bull, the image of the Bull Star MUL.GU$_4$.AN.NA, is the representation of the word "king," whereas the other prisms use the Lion, the image of the constellation Leo, normally written MUL.UR.GU.LA or MUL.LUGAL as the writing of "king." The reading of the glyphs on the Stone Prism, following Michael Roaf and Annette Zgoll's treatment, is "Esarhaddon, king of the land of Assyria, king of the four quarters" (*Aššur-aha-iddina šar māt Aššur šar kibrāt arba'i*, written Aššur-ŠEŠ-SUM GU$_4$ KUR Aš-šur GIŠIMARRU UB.DA.LIMMU$_2$).

In the array of symbols used for the representation of the royal name in glyphic form, the images with astral identifications are as follows: On the prisms is the lion, or MUL.LUGAL (*Šarru* "King" or Regulus, the brightest star of the constellation Leo, α Leonis, see Reiner and Pingree 1981, 13). On the Stone Prism of Esarhaddon, in place of the lion is MUL.GU$_4$.AN.NA (Alû "Bull of Heaven" or Is lê "Jaw of the Bull." This constellation includes the brightest star of Taurus, α Tauri (Aldebaran) plus the Hyades (Reiner and Pingree 1981, 12). The image of the bull is a well-known epithet (*rīmu*, written GU$_4$ or AM) for kings, gods, and heroes. The other two constellations are MUL.APIN or *Epinnu* "Plow," whose principal star is the brightest star of Triangulum, or α Trianguli (Reiner and Pingree 1981, 10) and MUL.AŠ.GÁN or *Ikû* "Field," consisting of the four brightest stars of Pegasus, namely α, β, γ Pegasi + α Andromedae or "The Square" of Pegasus (Reiner and Pingree 1981, 11). The image of the Field Star, or Square of Pegasus, here represents the concept of UB.DA.LIMMU$_2$, "four quarters," or "totality" by virtue of its sum of its four parts.

Correspondence is employed in other respects here too. Noteworthy is that between stars and gods so that in Esarhaddon's astroglyphs a connection can be made between the stars chosen to write the royal name and the deities they represent. In *Enūma Eliš* V 1–2, Marduk establishes "the stations" or celestial positions (*manzāzu*)[12] of the great gods and the constellations (*lumāšu*):

Enūma Eliš v 1–2 He created the (celestial) places for the great gods to stand; the stars, their (the gods') likenesses, he set up (as) constellations.

The compendium "Plow Star," MUL.APIN (Hunger and Steele 2019), is so named because it began its list of the stars of the "paths" of Enlil, Anu, and Ea with the Plow Star MUL.APIN. The astronomical compendium states the divine names that correspond to the constellations appearing as astroglyphs on Esarhaddon's inscription:

MUL.APIN I i 1 The Plow, (which is) Enlil, who goes at the front of the stars of Enlil
MUL.APIN I ii 1 The Bull of Heaven, the Jaw of the Bull, (which is) the crown of Anu
MUL.APIN I i 40 The Field, (which is) the seat of Ea, which goes at the front of the stars of Anu

(Hunger and Steele 2019, 29, 39, and 38)

Therefore, the images of constellations chosen for the astroglyphic writing of Esarhaddon's name and titles correspond to, and thus are the representations of, the great gods Enlil, Anu, and Ea, whose stars

[12] In an astronomical sense, *manzāzu* takes on the meaning of celestial position, but the basic meaning is a "place where (something) stands," as reflected in its logogram KI.GUB.

occupy the three main "paths" in the sky from north to south and therefore represent the entirety of the heavens. The Enlil stars include the circumpolar stars, the Anu stars are those of the middle region of the sky, and the southernmost stars belong to Ea.

Another important correspondence can be understood from Esarhaddon's inscription (Leichty 2011, 198, see p. 196), in which Marduk's Temple Esagil, which Esarhaddon restored in Babylon, was the replica (*tamšīlu*) of Ešarra and also the image/correspondence (*tamšīlu*) of the constellation MUL.AŠ.GÁN "the Field" (*Ikû*). Drawing upon a wide scope of material from *ṭupšarrūtu*, the Field star's appearance as an astroglyph evokes MUL.APIN's identification of *Ikû* with the seat of Ea. This constellation (the square of Pegasus) as the image of the "four quarters" was for that reason used for the writing of Esarhaddon's royal title referring to "the four quarters" or entire inhabited world.

Further to the astroglyphic use of the constellation The Field, correspondence was made between that constellation and the restored Babylonian Temple Esagil, whose foundations were laid as a square mirroring the Field Star. This idea is stated in a later text that verbally describes constellations and identifies the Field constellation not only with Babylon, but also with that city's New Year festival in the month of Nisan, the propitious start of the New Year:

> The Field, the seat of the god Anu, the one who goes at the front of the Stars of Anu, the month of [Ni]san, the New Year's Festival, the seat of the circle of the four quarters, the star of the New Year. [4 stars] are drawn side by side.
>
> (Uranology Text D Section A i 1–2, Beaulieu et al. 2018, 39)

The description of four stars "drawn side by side" is a perfect visual analog to what appears as an astroglyph. While they do not identify the image of the Field constellation in the Esarhaddon astroglyphs, Beaulieu et al. (2018, 4) note that it was a traditional method to represent stars as dots, and constellations as dots connected by lines, also seen in the so-called Assyrian planisphere (Koch 1989). The description of the Field, moreover, refers to it as the "circle of the four quarters" (*kippat erbetti*), evoking the image of the logogram ŠÁR. As a logogram, ŠÁR is not used for *kippatu*, but rather *kiššatu*, to mean "all of x," and its pictographic origins were a circle. When written with wedges, ŠÁR was rendered as a squared-off circle (see Figure 6.1).

Finally, with respect to the use of the Field star as an astroglyph, the knowledge of astronomical rules for regulating the calendar by the risings of constellations is embedded in the inscription. As stated in the Diviner's Manual (Williams 2002, 477):

Twelve are the months of the year. Three hundred and sixty are its days. Consider (lit.: take into our hand) the expected time of the *zagmukku* (beginning of the year). Investigate thoroughly the day of disappearance of the moon, the dates of the first appearances of the stars, their time periods, the correspondence (*mithurtu*) of the beginning of the year and the (constellation) Field.

(Oppenheim 1974, 200: 57–59; translation modified)

The intercalary rule, based upon the correspondence of the Field's heliacal rising with the New Year, draws the correspondence between the most propitious date in the year and the celestial writing of Esarhaddon's name and title.

As to the other of Esarhaddon's astroglyphic images mentioned in the uranology corpus, the bull MUL.GU$_4$.AN.NA/Alû/*Is lê* "Bull of Heaven/Jaw of the Bull" is mentioned and described (Text E rev. 15′) simply as "a bull." Visual analogy can be seen in the representation of the constellation *Epinnu* "Plow," which contains the V-shaped part of the constellation that we today still identify as Triangulum. It further bears a visual analogy to the signs AŠ + ŠUR, hence the use of the Plow to write the element "Aššur" in the phrase "Land of Assyria" (*māt Aššur*).

An additional example of the implementation of correspondence, but also indeed of nextness, are correspondences expressed numerically. In a number of tablets dating to the pre-Seleucid period, or between the sixth and fourth centuries, one method of constructing numerical relationships of correspondence was regularized to the extent that it appears in a group of table texts now referred to as calendar texts (*Kalendertexte*, Steele 2015, 188–90; Brack-Bernsen 2021). Lis Brack-Bernsen and John Steele (2004, 118) coined the termed "mathemagics" for the characteristic manner these texts exhibit in constructing a scheme for finding the astrological value of a date or a position in the zodiac. The following is a group of lines from a calendar text where the columns represent zodiacal sign, degree, month, and day in the ideal year:

10 7 1 1
7 14 1 2
4 21 1 3
1 28 1 4
11 5 1 5

(BM 96258 obv. 1–5; see Brack-Bernsen and Steele 2004, 96)

The numbers in column I represent zodiacal signs (the first entry "10" is Capricorn, for example) and decrease each line by three signs (modulo 12). The numbers in column II represent degrees and increase each line by seven (modulo 30). The "distance" from one zodiacal position from line to line is 277°, for example, Capricorn 7 to Libra 14 = 277°; Libra 14 to Cancer 21 = 277°; and so on. The numbers in columns III and IV can be interpreted as months and days in the ideal calendar, where a month always has thirty days. Thus, a correlation can be made between a position in the zodiac, such as Capricorn 7° (in line 1), and a "date" (month 1, day 1) in the ideal year, or vice versa. The equivalence of the twelve 30° zodiacal signs with the twelve thirty-day months in the ideal year was a basic device in Babylonian astrological texts that enabled such calculations, not based on actual but rather ideal astronomical relationships, to provide further positions for astrological interpretation.

A great many other examples make correspondence arguably one, if not the, principal theme running through much cuneiform knowledge. It was propagated throughout *ṭupšarrūtu* in omens, explanatory word lists, medical texts and commentaries, and in the tabulated calculations of lunar and planetary phenomena and in astrological schemata. It was once commonly held that lists were the chief shape of cuneiform knowledge, but what gave purpose and structure to the cuneiform word lists and omen lists, as well as to astronomical and astrological numerical tables, can be seen in the themes discussed in the foregoing sections, namely nextness, correspondence, and correlation between words, signs, numbers, and things comprising the epistemic and ontological world of the texts in the traditions of *ṭupšarrūtu*.

It must then finally be asked why the scribes, whose inquiries were guided, not only by epistemological but also by ontological commitments about the nature of their world, held such an abiding interest in correspondences among so many elements of that world: words, ways of writing words, numerical relations, concrete and imagined phenomena, and the relation of world phenomena to human beings. As in the case of nextness, which seems to have functioned as a nonphysical ordering property of several classes of phenomena (celestial, terrestrial, societal), and as such operated as an ontological feature of those phenomena, so too does the relation of correspondence seem to function as an ontological feature, a category by means of which to understand the nature of things in the world. The corpora of *ṭupšarrūtu* therefore contain evidence of two clear ontological categories, property and relation. Both property and relation are a part of the ontological classification system developed in post-Aristotelian Western metaphysics (Hoffman and Rosenkrantz 1997), but in relation to a way of worldmaking foreign to that which emerges from the cuneiform texts.

The following final section explores proportion and measurement as another possible theme and instrument of worldmaking.

PROPORTION AND MEASUREMENT

The role of numbers and measures in imagining and describing the world in cuneiform texts is evident from the Old Babylonian period and persists until the end of the tradition of cuneiform writing. As such, it seems to be another theme of cuneiform worldmaking. Early texts testify to the practical purposes of mathematics, such as the calculation of the areas of fields, the volume of granaries and storerooms, or the dimension of canal trenches. Mathematical relationships came to be of interest, as evidenced in the two primary Old Babylonian mathematical text types, that is, the list, or table text, and the problem text. Problem texts cover subjects sometimes related to real situations, sometimes rather as exemplifications of procedures for solutions. Some are terse statements of a problem, for example, "The side of a square is 1 UŠ. I made a border each side and I drew a circle. What is its area?" (BM 15285; see Robson 1999, 209; a schoolboy's tablet with this exercise can be seen in Friberg 2008, figure 11).

Other problem texts offer worked examples, such as

> [A heap. The circumference is 0;30, the height 1.] In 1 cubit, [what is the slope? You:] Take the reciprocal of 1, the height. You will see 1. [Multiply 1 by 0;30, the circumference.] You will see 0;30. Break 0;30 in half. You will see [0;15]. The slope is 0;15 in 1 [cubit. This is] the method.
>
> (BM 96954+, see Robson 1999, 224 xx 35–38)

The table texts list numerical relationships such as reciprocals (Aaboe 1964, 10–16), Pythagorean triples (exemplified by the Plimpton Table, see Aaboe 1964, 25–27 and Britton, Proust, and Shnider 2011), or various coefficients termed IGI.GUB/*igigubbûm* – sometimes translated as constants key numbers or constant factors.

Coefficients were used to calculate variable quantities in many diverse contexts. As an example from the realm of administrative accounting, coefficients were established for calculating the volume of storage of a variety of materials. The following is a compilation from several coefficient lists of various storage coefficients:

A storehouse: 0;3,36 is its coefficient.
0;3,36, of the storage of straw.

15, of the storage of water.
50, a cargo-boat of reeds.

(Robson 1999, 122 §7.4.1)

Coefficients were used in problem texts, as in the following solution to the problem of how many bricks are needed for construction of a wall:

[A wall.] The width is [2 cubits], the length 2 ½ nindan, the height is 1 ½ nindan. [How many bricks? You]: multiply 2 cubits, the width, by 2 ½ nindan, the length. You will see 0;25 (the area). Multiply [0;25] by 18, the height. You will see 7;30 (7 ½ sar, the volume). Multiply 7;30 by 6, the coefficient of a wall. You will see 45. The bricks are 45 sar. The method.

(BM 96957+, see Robson 1999, 232 i 1–4)

A scholarly compendium known as i_5-na_8 GIŠ.HUR.MEŠ AN *u* KI (BM 47860, see Livingstone 1986, 17–23)[13] includes many coefficients for the volume of storage units in the main temple of Nippur. For example, in a late source dated to the thirty-third year of Darius the Great (488 BCE), 4,20,6,40 is given as the coefficient (*igigubbû*, written IGI.GUB-*ú*) for the volume of the Amamma granary. This is followed by other coefficients for the volumes of various storerooms of the Ekur temple (Livingstone 1986, 20–21).

In the context of astronomy, coefficients quickly became an important mathematical tool used in the description and modeling of celestial phenomena. Coefficients used for astronomical calculation of the length of daylight and the duration of lunar visibility were included in the lists devoted to coefficients referred to before. The following appear in such coefficient lists:

12, the water-clock
3, 45, the visibility (of the moon)
40, to change day into night

(Robson 1999, 129 §8.2.1)

The coefficients given in the previous examples are stated without corresponding units of measure, but for measuring and assessing numerical relationships between things a complex metrology grew up alongside early mathematics.

[13] The title of this text series, as, suggested by Strahil Panayotov, refers to the "'eyes (*īnā*) of the plan of heaven and earth', with 'eyes' being a common metaphor in all Semitic languages for a 'spring' or 'source,'" and is parallel to two other expressions in KAR 44: 30–31: *kullat nagbi nēmeqi* and *pirišti lalgar*, referring to sources or springs of secret or esoteric knowledge. Both accord well with the idea of "the 'sources' of the plans of the universe" (*īnā uṣurāt šamê u erṣeti*), for which see Geller 2018, 308, note to line 31.

A significant example of the early necessity to establish such relationships was the creation of calendars in which months and days needed to be counted in order to define the year. In the Sumerian mythological literary text "Enki and the World Order," the god Enki was praised as the one who made the days and months to complete the year and thereby establish order in the world:

Counting the days and putting the months in their houses, so as to complete the years and to submit the completed years to the assembly for a decision, taking decisions to regularize the days: Father Enki, you are the king of the assembled people!

(ETCSL project 2006, ETCSL t.1.1.3.17–20)

Sexagesimal units of reckoning homogeneous time (Rochberg-Halton 1989, 146–50; Brack-Bernsen 2007), originally devised for accounting purposes, entered into early Babylonian astronomy as well as into the tradition of *Enūma Anu Enlil* omens, to remain in the cuneiform scholastic tradition for two millennia. These are the *bēru* "double-hour" and UŠ "degree," where 12 *bēru* = 1 day, 1 day = 360 UŠ, and 1 UŠ = 4 minutes of time. Local Sumerian calendars in the Ur III period had variable month lengths, but calendar months in the city of Nippur came into broader use. They were thereby standardized as the month names of the ideal calendar (twelve thirty-day months = one ideal year) common to the scholarly traditions of astrology (celestial and natal divination) and astronomy (MUL.APIN, Astrolabes) prior to circa 500 BCE and even later in some cases. The ideal calendar survived in Late Babylonian astrological texts, for example in the so-called calendar texts (*Kalendertexte*, Weidner 1967, 41–49, 50–52; Brack-Bernsen and Steele 2004; Steele 2015; Brack-Bernsen 2021 and 2022).

Weight measures were also incorporated into texts dealing, for example, with the length of daylight and may have reflected the use of a water clock with which the weight of water measured in mina indicated a length of time (Brown et al. 1999/2000): "1½ minas weight (of water in the water clock) is (equal to) 9 degrees linear distance" and see CAD, s.v. *manû* A meaning 3 b "mina," where 1 mina = $^{1}/_{60}$ of a talent). In early Babylonian astronomical texts, where units are sometimes not expressed, either UŠ or mina can be interpreted, where 1,0 (60) UŠ = 1 mina. For observational purposes, units familiar from the texts of land surveyors, such as the *ammatu* cubit or "forearm" and *ubānu* finger, are also found in astronomical contexts where observational distance between celestial bodies is of interest (Steele 2007b).

Measurement of world parts enhanced the mathematical proportions of complementary pairs, such as Marduk's measurement of the size of the world regions of Apsû, Ešarra, and Ešgalla in order to make the world proportionate. Proportions were built into the early schemes for describing daylight length variation through the year, as attested in the scheme of the Astrolabes and MUL.APIN, where 2:1 is the ratio of longest to shortest day, and 1/15 the length of night is the variable daily retardation of the appearance of the moon in *Enūma Anu Enlil* Tablet 14. Period relations generally are about proportions: in Neugebauer's terminology, the relation between the numbers he called Π and Z (discussed in Chapter 5), which construct the return of planetary synodic appearances to their start point with respect to a degree in the zodiac or to a date expressed in tithis, is another inflection of the theme of proportion.

Apart from astronomical texts as such, measurements and norms (mean of measurements) in relation to parts of the world emerge in scholarly texts as having a certain thematic importance. Divination employed the sense of a norm, not always determined by mathematical measurement, or even by number, but such norms were the means by which a phenomenon could be judged to be normal or abnormal, thus portending good or ill (Brown 2000, esp. 113–22). That is, what was normal could be gauged in terms either of a mean of measurements or a nonnumerical ideal, where that ideal is determined by the divine scheme of things (in relation to the scheme for the length of the day, for example, see Brown et al. 1999/2000, 137–38).

The idea of order by measurement, and what is considered numerically ideal (Brown 2000, 106), manifests itself as well in cosmogonic mythology and astrology. In the corpus of mythological literary texts, measurement can be found in a number of texts describing parts of the world. As in *Enūma Eliš*, Marduk measured the size of the world regions of Apsû, Ešarra, and Ešgalla in order to make the world proportionate (*Enūma Eliš* IV 141–46, Lambert 2013, 95). In the account of the rebuilding of Marduk's temple in Babylon by the Assyrian king Esarhaddon, deliberate reference is made to the Babylonian *Enūma Eliš*, repeating that Marduk's temple Esagil is the counterpart of the Apsû (*maṭṭalāt apsî*) and the equivalent, or likeness (*tamšīlu*), of Ešarra. The king conveys the care he took in restoring this temple by emphasizing the precise measurement of its dimensions and their complementarity to structures in the divine realm, including the constellation Field (our Square of Pegasus):

[In] a favorable month, on a propitious day, I laid its (Esagil's) foundation platform over its previous foundations (and) in exact accordance with its earlier plan I did not diminish (it) by one cubit nor increase (it) by half a cubit. I built and completed Esagil, the palace of the gods, an image (*maṭlāt*) of the Apsû, a replica (*tamšīl*) of Ešarra, a likeness (*mehret*) of the abode of the god Ea, (and) a replica of (the square of the constellation) The Field; I had (Esagil) ingen[iously] built (and) I laid out (its) square (*mithartu*).[14]

(Leichty 2011, 198 Esarhaddon 104 iii 41b–iv 1, and 206 Esarhaddon 105 iv 37b–v15)

Another kind of measurement and its role in establishing proportion in the world is expressed in scholastic and literary texts, for example, in the ritual explanatory text KAR 307 (Livingstone 1989, 99–105, nr. 39), which contains a curious passage that may be relevant here. Following its explanation of the Tigris and Euphrates as the right and left eyes of Tiamat (KAR 307 rev. 3), the text says, "40 *bēru* is the disk/circumference (variously read *sihirtu*, or simply HAB-rat) of the sun (Šamaš); 60 *bēru* is the disk/circumference of the moon (Sin)." The same statement where the sun's disk(?) is forty and the moon's is sixty appears in *Enūma Anu Enlil* Venus omens (Virolleaud 1908, Ištar 28:42). Similar measurements in mythological contexts are, for example, in the so-called Labbu myth where the monster Labbu is described:

Its length was 50 *bēru* [its breadth] 1 *bēru*.
Its mouth was 6 cubits, [its . . .] 12 cubits.
Its [ear]-flaps were 12 cubits.
For 60 cubits it [. . .] birds.
For 9 cubits it drew along [. . .] in the water.

(Lambert 2013, 364–65, CT 13 34–35:8–12)

Another kind of numerical relationship used in astrological schemata found its way into the manner in which planetary positions in the zodiac at the moment of birth could have a relation to one another. Greco-Latin astrology later termed these "aspects" (*aspectus*) of one planet to another, such as trine (120° separation), opposition (180° separation), quartile (90° separation), and sextile (60° separation). There is by now clear

[14] Measurement of the dimensions of temples and city walls is carefully detailed on tablets from the seventh century focused on the Esagil and Ezida (VAT 9961 + 10335; George 1992, 120–29) and from Seleucid Uruk, the famous Esagil Tablet (AO 6555 and BM 40813; George 1992, 109–19).

evidence of trine and quartile aspect in cuneiform texts. The trine relationship is attested in a Late Babylonian eclipse omen text (BM 36746 obv. 5′–6′ and 10′–11′; Rochberg-Halton 1984, 136):

> If the moon is eclipsed in Leo and finishes the watch (of night) and the north wind blows; Jupiter does not stand (in) the eclipse; Saturn and Mars stand in Aries or in Sagittarius. . . .
>
> If the moon is eclipsed in Virgo and [finishes the watch (of night)] and the south and east winds blow; Venus is not visible in its eclipse . . . Saturn and Mars . . . in Taurus or in Capricorn are visible.

This excerpt shows that a relationship was schematized for the position of the eclipsed moon in the case where the benefic planets, Jupiter or Venus, are not visible in the moon's own sign, but the positions of the malefic planets, Saturn and Mars, are located in signs 120° from that of the eclipsed moon. In the quoted excerpt, the first configuration has the eclipsed moon in Leo, the malefics in Aries and Sagittarius, thus forming a "trigon" (τρίγωνον "triangle"), hence "trine" aspect where each sign is 120° apart. The second example has the eclipsed moon in Virgo with the malefics in trine aspect in their positions in Taurus and Capricorn, respectively. The Babylonian system essentially made four groups of three signs, much as four groups of three months was a standard scheme in celestial omen divination (Rochberg-Halton 1984, 121–22, 129; Rochberg-Halton 1988b, 60–61).

The quartile aspect, or three groups of four quadruplicities, has been identified as well in the contexts of Hellenistic Babylonian astrology and medicine (Rumor 2021; Brack-Bernsen 2022). What came to represent geometrical relationships for Greek astrology were therefore rooted in the thematics of proportion and measurement characteristic of Babylonian worldmaking. In the case of astrological aspect, the Babylonian forms were not to be found in spherical geometry, but in the practice of making symmetrical and proportional arrangements of elements into groups, such as four groups of three to form a "triplicity," and three groups of four to form a "quadruplicity." Such practices go back to seventh-century BCE celestial divination and the division of the ideal year into four groups of three months to correlate with the four winds, or "directions," of the lunar eclipse shadow in terms of the quadrants of the moon (Rochberg-Halton 1984, 128–29; 1988a, 51–55).

7

Imago Mundi

> Correctness of representation like correctness of description varies with system or framework; the question "Is the picture correct?" is in this way like the question "Does the earth move?"
>
> – Nelson Goodman, 1978

The discussion of instruments of worldmaking in Chapters 5 and 6, especially the interest of Chapter 6 in the topographical aspect of the Babylonian world and its component parts, set the stage for the present chapter which considers the question of whether a specific geometrical shape can be inferred for the cuneiform scribes' (or any cuneiform text's) conception of the heavens, or indeed of the world as a whole. As identified in Chapter 6, some consistent thematics of cuneiform worldmaking emerge in terms of the particular topography of world parts and the relationships between them in the Assyro-Babylonian/Babylonian context. "World," as we would say, was expressed as the merism of heaven and earth, a duality from which themes necessitating a relationship between two parts are derived. Even in the source KAR 307 (see p. 171) referring to upper, middle, and lower heavens, the themes of counterparts and complementarities are present because the three levels of heaven were complemented by their counterpart upper, middle, and lower levels of earth. By means of this arrangement, the world was envisioned whole and complete.

A different theme, not requiring the dualistic conception "heaven and earth," concerned measurement and proportion, as discussed in the previous chapter. These themes rarely refer to the heavens alone. A single exception is found in a source discussed on p. 205, in which the circumference of one of the celestial circles (the Path of Enlil) is measured by a special unit devised for the purpose.

None of the available sources make it possible even by inference to give an affirmative answer to the question of whether the Babylonian world in its totality, or even of the heavens alone, was conceived of in geometrical terms, specifically as a sphere. It might seem to be a relevant question given that astronomy has been central to this book, but indeed it is a conception belonging to a wholly different historical and cultural-intellectual framework, namely that of ancient Greek philosophy. To facilitate the explanation of the positions of the moon and planets from an observer's position on earth, we often appeal to the idea of the celestial sphere *as a model*. For the purpose of introducing the basic celestial mechanics necessary for understanding the kinds of visible phenomena calculated by the Babylonian astronomers, Asger Aaboe said:

> We have no simple means of judging distances to celestial bodies; we can only determine the directions toward them. To describe what we thus observe, we introduce *the celestial sphere* as a spherical surface with its center at the observer's eye and with unit radius – which unit is irrelevant, but it is often thought more comfortable to choose a very large one. A celestial object is then mapped at, or identified with, the point on this spherical surface at which the line of sight to the object pierces the celestial sphere. The study of the behavior of this map of celestial objects on the celestial sphere is called spherical astronomy.
>
> (Aaboe 2001, 1–2; emphasis in the original)

Aaboe made clear that the celestial sphere was a device introduced for convenience and clarity of explication to a modern audience, not as a historical interpretation. More recently, John Steele (2007b, 306–20) outlined the defining features of the Babylonian astronomical framework, emphasizing the absence of a spherical conception. The fact that the spherical model so effectively represents how the celestial bodies move through the heavens and facilitates the prediction of their positions has no bearing on its use in Babylonian astronomical theory, or indeed in mythological or other accounts found in literary texts.

A spherical world picture finds no purchase in cuneiform sources, nor are spheres to be found in Babylonian mathematical texts. In all the algebraic geometry of the cuneiform mathematical texts, in which diagonals of rectangles and the areas of triangles and squares, chords of circles, and the areas of irregular polygons of "fields" are calculated (Friberg 2007, 202–19; Hoyrup 1990, 2017, and throughout), there are no spheres. In this context, it is, therefore, important to consider Piotr

Steinkeller's proposal (2005, 18) to envision the cosmological setting of the nocturnal extispicy ritual (*ikrib mūšiti*) as a spherical one.

The extispicy ritual with its accompanying prayer (*ikribu*) was performed to solicit a divine judgment from the sun god, Šamaš, whose judgment was expected to come in the form of signs the god would write upon the liver of the sacrificed lamb, thereafter to be inspected by the *bārû*-diviner and read as ominous. In these extispicy prayers and related texts, as Steinkeller pointed out, the sun god traversed the heavens during the day and descended into the netherworld, the realm of the dead, at night to render judgment about the cases for which the nocturnal extispicy ritual was performed. The stars, gods of the night, stood watch in heaven as advocates for the poor soul awaiting judgment. At the beginning and end of each day, the sun god and the gods of the night entered and/or exited the gates to the netherworld to serve their designated purpose for humanity.

This scenario enacts one of the important relationships the Babylonian scholars constructed between the divine and human realms. Steinkeller concluded:

> The Babylonians believed the act of extispicy to be a cosmic event in which both halves of the universe are actively involved. In the center of this event rests the sacrificial lamb. The lamb is transfixed with an axis, at one end of which stand the sun god and other divine judges, and at the other end, the "god of the night." The judges conduct a nocturnal trial over the person on whose behalf the extispicy is performed; the "god of the night" acts as his defender and protector. The decision is made essentially by the sun god, owing to his ability to see the future. It is then duly recorded in the intestines of the lamb. Adad, the wind, may be responsible for the delivery of the *ikribu* rituals and for the maintaining of communication between the two poles of the event more generally.
>
> (Steinkeller 2005, 46)

This is a compelling reconstruction of the imagined cosmological topography underlying the extispicy prayers. His summary statement states in addition:

> Most importantly, the universe was thought to be spherical. More specifically, it was composed of two complementary hemispheres: the upper hemisphere – or "upper parts" (elâti); and the lower hemisphere – or "lower parts" (šaplâti). In this scheme the two hemispheres are virtually identical counterparts of one another: each of them has its "earth" (ki, erṣetu) and its "sky" (an, šamû), the only difference being that the upper hemisphere is inhabited by the living, while the lower one is the domain of the dead.
>
> (Steinkeller 2005, 18–19; my emphasis)

Two equal halves of the cosmos forming the complementarity of heaven (*šamû*) and earth (*erṣetu*) seems to me less of a geometrical image and more of an expression of the theme of complementarity, as discussed in Chapter 6. Steinkeller cites Heimpel, who was less committal about the sphericity of the world, and who concluded his own reconstruction with the following statement:

> Possibly the most significant result of the present study is the establishment of the concept of heaven's interior as the invisible part of heaven below the earth. This concept implies the belief that the sun and the other heavenly bodies remained at all times in heaven and did not pass into the realm of the earth upon setting. The concept of heaven's interior in combination with the concept of the restless sun may be seen as important in the evolution of the recognition of the spherical shape of the heavens. There exists, for the time being, no indication that such recognition occurred within the frame of Babylonian culture.
>
> (Heimpel 1986, 151)

Therefore, Heimpel trod lightly on the possibility of the spherical image of the heavens as enveloping the earth of our waking existence, and he also (1986, 151, note 62) cited Neugebauer's disavowal of the spherical universe from the perspective of astronomical texts. Neugebauer said the technical analysis of the Babylonian astronomical texts that concerned him in Book II of his *History of Ancient Mathematical Astronomy*, the book devoted to Babylonian astronomy, indicated to him "the working of a theoretical astronomy which operates *without any model of a spherical universe*, without circular motions and all the other concepts which seemed 'a priori' necessary for the investigation of celestial phenomena" (1975, 348; my emphasis). The purpose of the present discussion, however, is less to argue against Steinkeller's and Heimpel's reconstruction of the spherical image of the world in cuneiform texts and more to draw attention to the intrinsic interest of the cuneiform *imago mundi* as expressed in the sources. As Heimpel said (1986, 150), the textual descriptions admit a number of contradictory conclusions, and regardless of any textual proof of one reconstruction or another, both Steinkeller and Heimpel were right to point to the significance of the extispicy material for any discussion of an *imago mundi* in cuneiform culture.

Further considerations are also worth noting, from both the mathematical and the astronomical sides. Because of their use of sexagesimal notation and metrology, the Babylonian scribes codified the 360° circle and utilized this concept for the zodiac and the unit UŠ "degree" of which there were thirty UŠ per sign. In a hymn to the sun god, Šamaš, the phrase "circle of heaven" (*kippat šamê*; 4R 19 nr. 2:51–52) is found, paralleled in a prayer to Šamaš and Adad with the phrase "circle of earth" (*kippat erṣeti*; Horowitz 1998, 274). In astronomical texts, *kippatu* has one attestation: in the phrase "the coefficient for the 'circle'" (*kippat ina igigubbû*; Neugebauer [1955], 1983 nr. 211 Sect. 2:5; revised in Ossendrijver 2012, 502, nr. 96: 5′). The meaning of *kippatu* in Late Babylonian astronomical procedures seems to relate to the calculation of eclipse magnitude (Ossendrijver 2012, 504, 596).

A sphere, on the other hand, necessitates geometrical concepts of radius and surface area, and the idea of equidistance from a center. All well attested in Greek geometry and philosophy, these ideas made deductions to the shape of the cosmos possible, most notably in Plato (*Tim.* 33b–34a) and Aristotle (*De Caelo* 2.4.286b10–287a2). The evidence points to the first appearance of the spherical cosmos in the fourth century BCE, probably tracing back to Parmenides' metaphysical reasoning in the poem on "What Is" (τὸ ἐόν[1]), from which there survives a text and quotations, some quite extensive, by Plato, Sextus Empiricus (second century CE), and Simplicius (sixth-century CE commentator). Further metaphysical arguments are given in Plato's *Timaeus* that the created world was "perfect and whole, with all its parts perfect" (Plato, *Tim.* 32d–33a, Bury 1929, 60–61) and in Aristotle's *On the Heavens* as the shape "most suitable to its substance, and is the primary shape in nature" (*De Caelo* 2.4.286b 10, Guthrie 1971, 154–55). Within the Greek spherical world picture, astronomy (from the fourth century BCE to the second century CE) constructed geometrical spherical models to describe the motions of the planets. The sphericity of earth was also determined by Aristotle to be a consequence of the physics of matter (*De Caelo* 2.14.297a8).

Between the circle and the sphere, however, is a world of ontological difference. It cannot be stressed too much that the spherical world was the product of Greek philosophical speculation, in which beauty was a central idea. This nonempirical notion was combined with the techniques of spherical geometry (and eventually trigonometry) for modeling the uniform circular rotation of heavenly bodies, sometimes referred to in Greek sources as "saving" the phenomena (Goldstein 1997; Jones 2016, 476). Greek astronomy sought to account for the observable synodic events in the heavens, which did not exhibit uniformity in their motion – the sun did not travel at constant speed throughout the year, and the planets did not

[1] Or, literally, "the being." But, as Alan Bowen explained to me (personal communication), "Ontology is usually cast as the study of being – which *may* be the same as the study of what there is and, further, *might be* the same as the study of what exists." In the leaps from one to another of these possibilities, as he pointed out, there is room for much difference of analysis of the meaning of the Greek τὸ ἐόν. My focus is not so much on "being," understood as what it is to be, as it is on what is or, better, what it is that is. It is in the sense of "what there is" that I have employed the term ontology throughout this book.

always move from west to east against the stars as they completed their synodic cycles – by appeal to rational explanation. The celestial phenomena were "saved" by the physical geometry of the cosmos, by means of which cinematic models could account for the difference between what we (merely) observe (say, retrograde motion) against the background of the fixed stars and what occurs behind the veil of perception, that is, by reason of the theory of uniform circular rotation around the geocentric sphere.

The cuneiform evidence does not support a concern with saving the phenomena; mathematical prediction of the phenomena, both the dates and the celestial positions of their appearances in the zodiac, was the goal, not the resolution of a distinction between mere appearances and an underlying true reality. One may well ask, however, whether the use of ideal periods and the ideal calendar is an inflection of such thinking. Important though such ideas of an ideal month, an ideal year, and an ideal mean of measurements (Brown 2000, 106, 113–22; Rochberg 2016a, 207–9) certainly were, translation to a distinction between appearance and reality such as we get from Parmenides (the "way of truth" and the "way of opinion") or from Plato (in the allegory of the cave) seems far-fetched. Babylonian scholars were more interested in problems of anomaly in accounting for nextness, and ways of resolving anomaly by a mathematical or conceptual means (Rochberg 2016a, 22–23, 108, 120–27) than by any such Parmenidean or Platonic ideas of appearance versus reality.

Instead of the celestial sphere, the horizon served as the reference point for the identification of stars and planets in early Babylonian astronomy. MUL.APIN, for example, named three paths corresponding roughly to intervals along the eastern horizon where certain fixed stars were observed to rise and follow their way across the sky. In Late Babylonian astronomy, another path referred to the moon's passage through the zodiacal circle (*mālak ^{d}Sin*), and was given a width of 12° (Neugebauer [1955] 1983, 190–91; ACT 200, obv. i 20; now Ossendrijver 2012, no. 53, 360–61). The sun too was seen as "displacing itself" (Akkadian *nasāhu*, CAD s.v. meaning 7)[2] through the zodiacal circle, as attested the expression ZI ^{d}UTU "displacement of the sun" (Neugebauer [1955] 1983, 194; ACT 200, obv. ii 5; now Ossendrijver 2012, no. 53, 362–63).The interest of Babylonian celestial science in lunar and planetary synodic appearances, as opposed to motion around the celestial sphere, accounts for the

[2] Note that CAD s.v. *nasāhu*, meaning 7 usage c in astronomical texts, translates "having a velocity."

quantification of the displacement of the body as it progressed around the zodiacal circle (where body is defined as on p. 154), and underscores the irrelevancy of spheres in Babylonian astronomy. Similarly, the synodic period of a planet was defined in Babylonian mathematical astronomy as the number of synodic arcs the planet will complete before it returns to the next same position in the sky in its synodic cycle, reckoned by degrees of the zodiac (return in celestial longitude); all of this was entirely independent of a spherical geometry.

Another telling indication of a lack of geocentricity is found in the sequential arrangement of the planets found in Late Babylonian astronomical (including horoscope) texts. In all contexts where the five planets are enumerated, the sequence Jupiter, Venus, Mercury, Saturn, and Mars (Rochberg-Halton 1988c) was canonical from an early date. This arrangement is at odds with the typically Greek geocentric and spherical model of the heavens, which organized the five planetary orbits around earth in the order of their sidereal rotations with respect to earth. Thus the order Mercury, Venus, Mars, Jupiter, and Saturn indicates that the five planetary spheres were "above" those of the moon and sun. The Babylonian order of the planets was based upon their attributes as gods with celestial divinatory significance, either positive or negative, and not to sidereal periods. Accordingly, Jupiter, Venus, and sometimes Mercury are benefic; Saturn, Mars, and sometimes Mercury are malefic. Mercury holds an ambiguous position between the two benefics and the two malefics. The brightness of the benefics probably conferred such positive associations, as the dimness of the malefics accounted for their negative associations. Old Babylonian references to the planet Mars as a destroyer of the herds (Rochberg-Halton 1988a, 235, note 2) suggests an early origin of this doctrine. In the Neo-Assyrian period, Jupiter was a herald of propitious omens for Esarhaddon (Leichty 2011, 106; Esarhaddon 48.55–57a; 197 Esarhaddon 104.ii.34–41a; 204 Esarhaddon 104.iii.10–18; 245 Esarhaddon 116, rev. 2–4). Late Babylonian astrological texts make those assignments explicit (Rochberg-Halton 1988b) using a schema whereby a favorable sign is indicated by a bright benefic planet or a dim malefic one, and an unfavorable sign is indicated by a bright malefic or a dim benefic.

It is doubtless of some interest in this context to point out that sources from outside the corpus of mathematical astronomy reflect speculation, not on the shape but on the great size of the heavens. A number of texts dating to the seventh century as well as from Hellenistic Babylonia concern linear distances between culminating (*ziqpu*) stars (Schaumberger

1952, 1955; Rochberg-Halton 1983; Horowitz 1994; Beaulieu 1999, 99; Breger 2022). In these texts, a unit of measure called "celestial *bēru*" (*bēru ina šamê*) is employed for the distances between the culminating stars, or what we could call the differences in right ascension between these stars (for right ascension, see Glossary of Astronomical Terms Cited).

The *bēru* is a metrological unit attested since the third millennium BCE for measuring distance as well as time (the time it takes to go a certain distance). The same unit *bēru* is used for the measurement of arc, where it refers to the twelfth part of a circle: 12 *bēru* make up the circle of the day, or 1 rotation of the heavens, and since each *bēru* is subdivided into 30 UŠ, or 30° (= 120 minutes of time), the circle of the day, or 12 *bēru*, consists of 360° (= 24 hours). The *ziqpu* star texts tallying the total of the distances from one culminating star to the next in a great circuit made use of a new and very large unit for celestial measurement, calling it the "celestial *bēru*" (Akkadian *bēru ina šamê* "*bēru* in the sky"). The celestial *bēru* stands in relation to a new terrestrial unit called "degree on the ground" (UŠ *ina qaqqari*) at a ratio of 1:1800, or 1:30,0 expressed in the sexagesimal system, and parallels the ratio of the regular *bēru* to UŠ of 1:30.

With such large units as the celestial *bēru*, the texts tally the circuit of (in this case) 26 culminating stars as 655,200 celestial *bēru* (expressed sexagesimally as 3,2,0,0).[3] The total distance is meant to represent the size of the circle of the sky where the culminating stars are seen to cross the meridian, or what the Babylonian scribes called the "Path of Enlil." The circumference of the Path of Enlil was measured by the combined distances from one culminating star to the next in the series of twenty-six (AO 6478; Schaumberger 1955; Horowitz 1998, 183–84), although some texts give only twenty-five.

As just noted, 1 celestial *bēru* = 30,0 (= 1,800) degrees on the ground, which is proportionally equivalent to the standard ratio of 1 *bēru* = 30 UŠ. Just as 12 *bēru* = 6,0 UŠ (= 360°), 12 celestial *bēru* = 6,0,0 (= 21,600) UŠ *ina qaqqari* "degrees on the ground," or 60 times as many celestial *bēru*. This is a strictly numerical expansion based on the sexagesimal numbers 30 and 60, given to account for the great distance conceived of for the circle of the Path of Enlil. Here numbers, not physics, are the medium for speculation.

The early astronomical compendium MUL.APIN (Hunger and Steele 2019) lists the culminating *ziqpu* stars as a means of orienting the observer

[3] Steele 2014, 131 shows that the twenty-six-star list is the exception to the more common list of twenty-five *ziqpu* stars. MUL.APIN had fourteen, so the list varied over time.

of heaven relative to the stars. Here, the observer stands with respect to a local horizon, not with respect to an overall geometry of the cosmos. MUL.APIN tells us, "It is the *ziqpu* stars that stand in the path of Enlil in the middle of the sky opposite the chest of the observer of the sky, and by means of this perspective, he observes the rising and setting of the stars at night" (MUL.APIN I iv 1–3, Hunger and Steele 2019, 61–2). This is a clear statement that the observer looking to the center line of heaven has the horizon points east and west on either side of him. With the observation of the rising and setting of stars, we might be tempted to assume that the Babylonian astronomers extrapolated from the visible arcs that the stars make overhead to the idea of a spherical heaven, but temptation is (usually) best resisted.

The last and most decisive evidence against the idea of a spherical heaven, or a spherical earth, are the Babylonian schemata developed for calculating the length of day based upon a ratio of longest to shortest day (either 2:1 or 3:2). In neither of the two traditions of calculation schemes for the change in the length of the day in cuneiform was consideration given to a difference in what we would now call geographical latitude (Neugebauer 1942, 255; 1963, 533). In the absence of that concept, the scribes could not, indeed did not, describe Babylon as a point on a spherical globe. It was not until Greek texts of the Hellenistic period that other ratios of longest to shortest day length were derived to account for other locations on a spherical earth, hence the seven *klimata* of Hellenistic astrology and astronomy (Neugebauer and van Hoesen 1959, 3–5) which adopted the parameter 3:2 for Babylon to apply to the second *klima*, also given as that of "Syria" (Neugebauer 1975, 729). The incorporation of the Babylonian value for the *klima* of Syria carried through to produce a distorted map of the eastern part of the world (Neugebauer 1945, 19), because the parameter 3:2 produces a latitude for Babylon of 35° instead of $32^1/_2$°. The Babylonian ratio M:m of 3:2, or indeed by the value of the constant M, was very likely the point of departure for the Hellenistic Greek system of defining geographical latitude in terms of the characteristic value of M or the ratio of longest to shortest length of the day for a given locality. It cannot be shown, however, that the Babylonian astronomers themselves had the idea. As Neugebauer said, "It is interesting to realize that this method [of using M or M:m] was not merely adopted by the Greeks for special latitudes like Babylon or Alexandria, but consistently expanded as soon as the discovery of the sphericity of the earth required the consideration of changing horizons" (1942, 262).

Taken as a whole the evidence is against the idea of a spherical cosmos.[4] The Babylonian ratios of longest to shortest day, whether 2:1 or 3:2, both refer to the local Babylonian situation. Celestial circles, however, are implied by the image of the horizon as a giant *tarbaṣu*, a circular cattle pen, keeping the stars and planets in the sky (Rochberg 2010c, 348–52). The Path of Enlil was given a circumference in celestial *bēru* units, so that it too was conceived of as a circle. The zodiac was originally constructed as a mathematical circle of 360°, with twelve signs of thirty UŠ each. The idea of the rising times of six zodiacal signs, that is, half the zodiacal circle, as adding up to the length of the day was an original Babylonian idea, but even the evidence for the idea of rising times does not argue for a spherical cosmos, or for the idea that earth occupies the geometrical center of an enclosing fixed star sphere.

The significant difference between Babylonian and Greek astronomy is that Greek planetary models with their eccentric and epicyclic circles within each planet's sphere, as we know from Ptolemy's *Planetary Hypotheses* (Goldstein 1967), were more than mathematical models to account for the planetary phenomena. More than predicting the phenomena, the celestial sphere represented the actual geometry of the geocentric cosmos and was conceived as having material, physical existence. Within the enclosing whole, the planets moved within their spherical shells composed of aether, each attaining a closest and a furthest distance from earth as they whirled around inside their individual spheres in accordance with the model for their motion. As the *Planetary Hypotheses* explains:

> We began our inquiry into the arrangement of the spheres with the determination, for each planet, of the ratio of its least distance to its greatest distance [from earth]. We then decided to set the sphere of each planet between the furthest distance of the sphere closer to earth, and the closest distance of the sphere further (from the earth). Let us assume that only the spheres of Mercury and Venus lie below the sphere of the Sun, but that the others do not. We have explained in the *Almagest* that the least distance of the Moon is 33 earth radii, and its greatest distance 64 earth radii, dropping fractions. Moreover, the least distance of the Sun is 1160 earth radii, and its greatest distance 1260.
>
> (Ptolemy, *Planetary Hypotheses* Book 1 Part 2 §3; Goldstein 1967, 7)

[4] Referring to the Sasanian period and the cultural ambit of Iran, Antonio Panaino (2021, 43–46) noted that Walter Henning (1942) revived a nineteenth-century idea that Pahlavi *spihr* was a loanword from Greek σφαῖρα "sphere." Panaino prefers to understand *spihr* as an adaptation, not a loanword that came into Iran via a Syriac intermediary. He remarks, "The assumption that an Indo-Iranian word might mean 'sphere,' not in a basic and primitive sense such as that of a 'ball' … but that of (geometrical cosmic or celestial) 'sphere' is totally implausible" (2021, 51). I find this comports with the lack of cuneiform evidence for such an image in the Babylonian context.

This text argues for the arrangement of the planetary spheres in their ascending order from earth. It speaks of the eccentricity of the motions of the moon and of Mercury in their spheres and contrasts these motions to that of the outermost fixed star sphere. Ptolemy then says, "The sphere nearest to universal motion is the sphere of the fixed stars which moves with a simple motion, resembling the motion of a firm body whose revolution in itself is eternally unchanging" (Book 1 Part 2 §3, Goldstein 1967, 7).

As explained in the *Planetary Hypotheses*, on the basis of the Aristotelian cosmos of interesting spheres with no empty space between them, one planet's greatest distance from earth would be the next planet's least distance, in geocentric order moon, Mercury, Venus, Mars, Jupiter, and Saturn, followed by the outermost and most eternally unchanging fixed star sphere. By establishing the distances from earth of the inner radius (least distance) and outer radius (greatest distance) attained by the planetary models, Ptolemy calculated the size of the cosmos as a whole, that is, the size of the fixed star sphere. For this purpose, he first conceived a unit $^1/_{60}$ of the radius of the deferent and with this unit determined the ratios of the planets' least to greatest distances from earth. From the series of least to greatest distances, Ptolemy calculated the size of the cosmos using the terrestrial radius as his unit.[5] This was possible from the time of the third century BCE due to Aristarchus' and Archimedes' interest in distances from earth to the sun and moon and the consequent interest in the size of the cosmos (Neugebauer 1975, 634).

Questions as to the size of the cosmos, or the distances from earth to the planets, do not present themselves in Babylonian mathematical astronomy. The measurement of the Path of Enlil in celestial *bēru* was not an assessment of the size of the cosmos as a whole. Instead, totality is expressed as the sum of parts, as in "all of x (things)," where the things totaled are lands, cities, and people, including in some contexts "all heaven and earth" (Rochberg 2021). The expression "all heaven and earth" belongs exclusively to divine epithets, as illustrated by Marduk's epithet as the "sage of all heaven and earth" (*apkal kiššat šamê u erṣeti*; Tallqvist 1938, 29) or Nabû's as "the one in charge of all heaven and earth" (*pāqid kiššat šamê u erṣeti*; Tallqvist 1938, 382). These deities, as

[5] *Planetary Hypotheses* 1 ii 3, refers to *Almagest* 5.15, where the distances to the moon and sun in terrestrial radii are found. Eratosthenes had given an estimate of the earth's circumference in the third century BCE.

well as the sun god Šamaš and moon god Sin, manifested themselves in the world as celestial bodies, whose chief attribute was radiance (*melammu*; Rochberg 2009b, 49–50 and throughout) and who sent ominous signs through their visible appearances. As such, the celestial bodies were not theorized by the scribes as a part of a physical system within a physical whole, but within a world of ever-changing but sometimes determinately predictable visible phenomena (our physical natural phenomena). The phenomena in turn were of interest for their mutual relationships and for signaling favorable or unfavorable fortunes for human life.

Finally, I have tried to articulate the ways in which the world descriptions attested in cuneiform texts represent a world *in its own right*, but I realize that many will still be asking: if the physical material world was not the foundational idea for the Babylonian version of its world, then what was? The stakes for understanding physical phenomena go up in the context of science so long as we limit the scope of scientific knowledge and practice to the physical sciences or define science as *physical* science.

The disciplines of *ṭupšarrūtu* focused on appearances and morphological characteristics of things in the world, from plants and stones to ominous phenomena to planetary and lunar synodic appearances to the diagnostic signs of illness. The Babylonian scribes observed and classified these phenomena and schematized them by descriptive features – light, dark, bright, dim, by colors and shapes, or by analogy to things they might resemble (Rochberg 2016a, 103–27). The observable features and phenomena of the heavenly bodies, the human body, the human face, the internal organs of the sheep, and many other things were all subject to such descriptions, and the observable features of things carried meaning as ominous signs, or as signs and symptoms of illness. At the outset of this book, I suggested that science can be defined as cultural activities resulting in systems of knowledge of what a particular community imagines and perceives its phenomenal world to be, which in turn depends upon how and why certain phenomena become objects of inquiry. In terms of that definition, appearances and signs were intrinsic, indeed central, aspects of Babylonian empirical and theoretical science.

It seems an unavoidable conclusion that ominous signs held primary significance for cuneiform knowledge and science because an inherent property of many phenomena was conceived in terms of nextness. This is incontrovertible and easy to demonstrate in the case of celestial bodies; the cuneiform sciences of celestial divination, astronomy, and astrology grew up around that idea. However overwhelming the evidence for the idea of periodic nextness in celestial phenomena was, the idea apparently carried over, at least in the Late Babylonian period, to rains, floods, comets, and earthquakes (Hunger 1976; for TU 11: 23–28, Brack-Bernsen and Hunger 2002, 11 and 16). The tablet TU 11§23 explicitly cites the goal year periods of the planets as the means by which "you make," that is, calculate rains and floods. The recording in the Diaries of market "prices" (in terms of equivalent amount for one shekel of silver "which was given in the land" *ša ina māti innadnu*), that is, "the going rate" of six commodities (barley, dates, mustard [*kasû*], cress [*sahlû*], sesame, and wool),[6] suggests there as well that the question of cyclical recurrence was in the background, as van der Spek noted (van der Spek, van Leeuwen, and van Zanden 2014, 6). Similarly for the Euphrates river levels, recorded in the Diaries following the quotes on market prices, the careful recording of rising and falling river levels and in how many days (Sachs and Hunger 1988, 34–36) was likely motivated by the same question of cyclical recurrence. All these seemingly unrelated data were collected and appended to the monthly paragraphs detailing the lunar and planetary phenomena observed regularly by the scribes and included in the Diaries. If indeed a property such as nextness was of interest to the scribes, these data can be seen as closely related. Attention to the regular periodic appearance of a wide variety of signs from early periods was surely the prerequisite to the systematic observation of celestial (and noncelestial) phenomena in the Diaries, and the eventual construction of models for predicting them, whether by goal year methods or by the calculation methods of Systems A, B, and their variants.

What is clear is that an interest in periodic phenomena is no guarantee that phenomena will come to be thought of as part of a material order, or that there will be a search for physical causes. The absence of attention to the physical nature of the world either in its parts or in its entirety suggests a fundamentally different conceptualization of the phenomenal world from that of Greek natural philosophy. Where there is no *phusis*, there is no physics, hence the lack of speculation on planetary motion as such or on the material substance of the stars such as of fire or aether and, consequently, on the composition, shape, and size of the cosmos as a whole.

Thus cuneiform knowledge was no incipient, embryonic, or rudimentary *natural philosophy*. It was a fully developed system of knowing in its

[6] The identification of *kasû* as mustard and *sahlû* as cress has been disputed and discussed in Geller 2000.

own right, in which the regularity and irregularity of phenomena operated within a classificatory framework focused on the meaning and interrelationship of phenomena as signs. As a result, it was the appearance of things, particularly their regularity, sequence, and nextness more than their physical nature, material essence, or substance that was the object of scribal inquiry and systematization. Of course, phenomena classified as physical or natural can also be viewed as signs, as was common throughout the premodern West. The fact that nature was not an ontological organizing framework in Babylonia, however, has to be central to our understanding of the world order of cuneiform culture, its *imago mundi*, and the place of astronomy within it.

The dominant effect of *phusis/natura* on astronomy, cosmology, medicine, and astrology from the Classical world forward signals a deep-seated discontinuity from the ancient Middle Eastern traditions. That shift may not be as obvious as that which underpinned Kuhn's idea of the paradigm shift, namely from the geocentric to the heliocentric world picture, or Koyré's formulation of the transition from the closed to the infinite universe. It was more perhaps more subtle, and it did not impede the transmission of scientific ideas from Babylonia to the West, given that astrology, the study of signs, and magic all continued into the sciences of the Middle Ages and early modern Europe, as they did in Islamic sciences. Once the idea of nature and physics gained a foothold in those later intellectual traditions, however, the legacy of the cuneiform scribes was transformed by new ways of worldmaking.

Conclusion

Two strands of intellectual revisionism were essential to the project of this book, which set out to connect our historiography of the cuneiform sciences to an anthropology of science. The first was the rise of pluralism within the history of science, most famously but not exclusively represented by Thomas Kuhn, and the second was a more recent development, the "quiet revolution" within anthropology of the ontological turn.

The tradition of the ontological turn itself was predicated on the dismantling of the following long entrenched commitments: the binary of nature and culture, the inviolability of the Western conception of nature, especially of its universality and of what Barry Stroud called "the charm of naturalism." At the opening of the Presidential Address before the Seventieth Annual Pacific Division Meeting of the American Philosophical Association in Seattle, Washington, April 5, 1996, he said, "the idea of 'nature,' or 'natural' objects or relations, or modes of investigation that are 'naturalistic,' has been applied more widely, at more different times and places, and for more different purposes than probably any other notion in the whole history of human thought" (Stroud 2004, 21). For those concerned with other ontologies, whether historians or anthropologists, the dominance of nature as the modern framework for worldmaking is challenging to circumvent.

In historical terms, nature may be many things; it has been conceived as an organism, as mind or intelligence, conceived with gods, without gods, as self-moving, self-causing, as a teleology, as process, as a machine, as life. Many ideas have been attributed to nature, making it endlessly redefinable. But regardless of the manifold natures of nature, it has long been viewed as the other half of a dichotomy, as not-culture, as a separate and objective realm apart. The ontological turn was built upon critiques of these commitments and required a new stance, both against what was famously called the Great Divide (nature/culture) and against Western naturalism as the dominant ontology for the study of science (Tola and dos Santos 2020, 163).

Closely related but prior to the ontological turn that took place in anthropology was the philosophy of ontological relativity, associated most particularly with the mid-twentieth-century philosophers Willard van Orman Quine and Nelson Goodman. No longer so often cited, their contributions remain in the background of even the most recent studies in which ontology is a central concern. Goodman's project of worldmaking has been my other primary influence, even while a number of his key claims remain problematic, primarily the version to world relationship and the way in which multiple existent worlds, made via science or art or other cultural forms, might overlap synchronically. Thus from the literature of the ontological turn, especially from anthropology but also from science and technology studies (STS) and the philosophy of science, many twisted threads[1] have been woven together for this book's historiographical reflections on worldmaking.

The present study turned to ontological questions about the phenomena the cuneiform scribes identified, classified, and theorized about in their known world. This was not because I thought it useful to replace epistemology with ontology, but rather that epistemology, especially in the context of a remote and distant science, requires an admission that what the scribes were interested to know, that is, the phenomena comprising the objects of their scientific knowledge, had an inextricable connection to ideas about what those phenomena were and how they belonged to an ordered world. For this reason, the philosophy of worldmaking and Goodman's particular brand of ontological relativity (or *evanescent* ontology), together with the wealth of ideas on these questions from philosophy, STS, and anthropology, were introduced as a heuristic for analysis and interpretation.

As the question of the objects of scientific inquiry in ancient Babylonia (and Assyria) is a matter both of epistemology and of ontology, an inquiry into ontology in cuneiform sources is what brought me to worldmaking and the ontological turn, as discussed throughout this book. Objects of

[1] This is an allusion to a passage in a colophon of the eleventh-century BCE Babylonian scribe and erudite Esagil-kin-apli, in which he used the expression GU.MEŠ GIL.MEŠ "twisted threads" for the many sources he wove together to form the medical diagnostic series *Sakikku*. See Finkel 1988, 149 B: 18′; Geller 2018, 47; Schmidtchen 2018, 317: 52.

knowledge exist within worlds, material, immaterial, and conceptual. To abstract the objects of knowledge from those worlds may seem useful in some contexts, such as in mathematical modeling, but the denotational purpose of mathematical description and the exemplifying purpose of, say, the Babylonian omen texts, point beyond themselves to some framework of reference for such descriptions, models, and exemplifications. If epistemological standards are relative to history, so too are the frames of reference for historical sciences and thus the norms of ontology.

The conception of nature tends to produce a singularity in the way of worldmaking constructed around it, allowing only one world to support many worldviews. The many worldviews approach preserves the "prepackaged, but unfortunately undiscoverable reality" (Goodman and Elgin 1988, 49) of the one and only world (usually defined as the one known by modern science) but sacrifices the reality and legitimacy of the actual worlds made in other places and in other times. I have found that claim to be more than limiting from a historical standpoint, and troublingly ethnocentric from a cultural one. Here, too, Goodman's philosophy of worldmaking, in removing the dualism of the natural and the conventional while seeing their interdependence, becomes a useful heuristic (Mitchell 1986, 115).

With regard to the ontologies of Indigenous peoples, Viveiros de Castro (2003, 18) said the Indigenous populations of the Amazon did not live in a sustained fantasy. Neither did the Babylonian scribes. Once the world is conceived from the perspective of others, assuming that is achievable, the ontological world of modern science appears less and less as a reasonable standard by which to understand other orderings of reality, whether from ethnographical or historical sources. And because historical and ethnographical sources attest to a plurality of sciences practiced in accordance with many ways of worldmaking, their study can profit from all the methodological approaches we have at our disposal, that is, the history of science, the philosophy of science, the sociology of science, and, not least, the anthropology of science.

References

Aaboe, Asger. 1955. "On the Babylonian Origin of Some Hipparchan Parameters." *Centaurus* 4: 122–25.

1964. "On Period Relations in Babylonian Astronomy." *Centaurus* 10: 213–31.

1980. "Observation and Theory in Babylonian Astronomy." *Centaurus* 24: 14–35.

2001. *Episodes from the Early History of Astronomy*. New York: Springer.

Agassi, Joseph. 1963. *Towards an Historiography of Science*. The Hague: Mouton.

Ajdukiewicz, Kazimierz. 1934. "Das Weltbild und die Begriffsapparatur." *Erkenntnis* 4: 259–87. Reprinted as "The World-Picture and the Conceptual Apparatus," in Ajdukiewicz 1978, 67–89.

1978. *Kazimierz Ajdukiewicz: The Scientific World-Perspective and Other Essays 1931–1963*, edited by Jerzy Giedymin, 35–65. Boston: D. Reidel Publishing Company.

Alberti, Benjamin, Severin Fowles, Martin Holbraad, Yvonne Marshall, and Christopher Witmore. 2011. "Worlds Otherwise: Archaeology, Anthropology and Ontological Difference." *Current Anthropology* 52: 896–912.

Allen, R. E. 1984. *The Dialogues of Plato*. New Haven, CT: Yale University Press.

Alvargonzález, David. 2013. "Is the History of Science Essentially Whiggish?" *History of Science* 51: 85–99.

Amin, Samir. 1989. *Eurocentrism*. New York: Monthly Review Press.

Aspers, Patrik. 2015. "Performing Ontology." *Social Studies of Science* 45: 449–53.

Augé, Marc. 2006. "The Near and the Elsewhere." In *Anthropology in Theory: Issues in Epistemology*, edited by Henrietta L. Moore and Todd Sanders, 587–97. Oxford: Blackwell.

Beaulieu, Paul-Alain. 1999. "The Babylonian Man in the Moon." *Journal of Cuneiform Studies* 51: 91–99.

Beaulieu, Paul-Alain, Eckart Frahm, Wayne Horowitz, and John Steele. 2018. *The Cuneiform Uranology Texts: Drawing the Constellations*. Philadelphia: American Philosophical Society Press.

Ben-Dov, Jonathan. 2008. *The Head of All Years: Astronomy and Calendars at Qumran in Their Ancient Context*. Leiden: Brill.

2014. "A Jewish Parapegma? Reading 1 Enoch 82 in Roman Egypt." In *Time, Astronomy, and Calendars in the Jewish Tradition*, edited by Sacha Stern and Charles Burnett, 1–25. Leiden: Brill.

2018. "Babylonian Astral Sciences in West-Semitic Sources: The Case of Qumran." In *The Interactions of Ancient Astral Science*, by David Brown, with contributions by Jonathan Ben-Dov, Harry Falk, Geoffrey Lloyd, et al., 138–86. Bremen: Hempen.

Ben-Dov, Jonathan and Seth Sanders. 2014. *Ancient Jewish Sciences and the History of Knowledge in Second Temple Literature*. New York: New York University Press and Institute for the Study of the Ancient World.

Benthall, Jonathan. 2021. "The Critique of Ethnocentrism in Retrospect." *Anthropology Today* 37: 20–22.

Berlin, Isaiah. 1982. *Against the Current: Essays in the History of Ideas*. Harmondsworth: Penguin Books.

Bernal, Martin. 1992. "Animadversions on the Origins of Western Science." *Isis* 83: 596–607.

Berthelot, Marcellin and Charles-Émile Ruelle. 1887–88. *Collection des anciens alchemistes grecs*. Paris: Georges Steinheil.

Bezold, Carl and Franz Boll. 1911. *Reflexe astrologischer Keilinschriften bei griechischen Schriftsteller*. Heidelberg: C. Winter.

Biggs, Robert D. 1971. "An Archaic Sumerian Version of the Kesh Temple Hymn from Tell Abū Ṣalābīkh." *Zeitschrift für Assyriologie* 61: 193–207.

1990. "Medizin." *Reallexikon der Assyriologie* und Vorderasiatischen Archäologie 3: 625–26.

Blackburn, Simon. 2005. *Truth: A Guide*. Oxford: Oxford University Press.

2010. "Conceptual Schemes." In *Relativism: A Contemporary Anthology*, edited by Michael Krausz, 164–82. New York: Columbia University Press. First published in *Truth: A Guide*, pp. 199–204.

Böck, Barbara. 2009. "On Medical Technology in Ancient Mesopotamia." In *Advances in Mesopotamian Medicine from Hammurabi to Hippocrates: Proceedings of the International Conference "Oeil malade et mauvais oeil,"* Collège de France, Paris, June 23, 2006, edited by Annie Attia and Gilles Buisson, 105–28. Leiden: Brill.

2011. "Sourcing, Organizing, and Administering Medicinal Ingredients." In *The Oxford Handbook of Cuneiform Culture*, edited by Karen Radner and Eleanor Robson, 690–705. Oxford: Oxford University Press.

2015. *Shaping Texts and Text Genres: On the Drug Lore of Babylonian Practitioners of Medicine*. Barcelona: Editorial Ausa.

Boghossian, Paul A. 2006. *Fear of Knowledge: Against Relativism and Constructivism*. Oxford: Clarendon Press.

Bohak, Gideon and Mark Geller. 2013. "Babylonian Astrology in the Cairo Genizah." In *Envisioning Judaism: Studies in Honor of Peter Schäfer on the Occasion of His Seventieth Birthday*, edited by Ra'anan S. Boustan, Klaus Herrmann, Reimund Leicht, Annette Yoshiko Reed, and Giuseppe Veltri, 607–22. Tübingen: Mohr Siebeck.

Borrelli, Noemi and Eduardo A. Escobar. 2022. "Crafting Purity in Assyro-Babylonian Procedures: Time, Space, and the Material World." *Arys. Antigüedad: religiones y sociedades* 20: 27–76.

Bowen, Alan C. 2007. "The Demarcation of Physical Theory and Astronomy by Geminus and Ptolemy." *Perspectives on Science* 15: 327–58.

2013. "Three Introductions to Celestial Science in the First Century BC." In *Writing Science: Medical and Mathematical Authorship in Ancient Greece*, edited by Markus Asper, 299–327. Berlin: De Gruyter.

Bowen, Alan C. and Francesca Rochberg, eds. 2020. *Hellenistic Astronomy: The Science in Its Contexts*. Leiden: Brill.

Boyarin, Daniel. 2009. "Rethinking Jewish Christianity: An Argument for Dismantling a Dubious Category." *Jewish Quarterly Review* 99: 7–36.

Brack-Bernsen, Lis. 1969. "On the Construction of Column B in System A of the Astronomical Cuneiform Texts." *Centaurus* 14: 23–28.

1980. "Some Investigations on the Ephemerides of the Babylonian Moon Texts, System A." *Centaurus* 24: 36–50.

1990. "On the Babylonian Lunar Theory A: Construction of Column Φ from Horizontal Observations." *Centaurus* 33: 39–56.

1997. *Zur Entstehung der babylonischen Mondtheorie: Beobachtung und theoretische Berechnung von Mondphasen*. Stuttgart: Steiner.

1999. "Goal Year Tablets: Lunar Data and Predictions." In *Ancient Astronomy and Celestial Divination*, edited by Noel M. Swerdlow, 149–77. Cambridge, MA: MIT Press.

2003. "The Path of the Moon, the Rising Points of the Sun, and the Oblique Great Circle on the Celestial Sphere." *Centaurus* 45: 16–31.

2005. "The 'Days in Excess' from MUL.APIN: On the 'First Intercalation' and 'Water Clock' Schemes from MUL.APIN." *Centaurus* 47: 1–29.

2007. "The 360-Day Year in Mesopotamia." In *Calendars and Years: Astronomy and Time in Ancient Mesopotamia*, edited by John M. Steele, 83–100. Oxford: Oxbow Books.

2010a. "Prediction of Days and Pattern of the Babylonian Lunar Six." *Archiv für Orientforschung* 52: 156–78.

2010b. "Methods for Understanding and Reconstructing Babylonian Predicting Rules." In *Writings of Early Scholars in the Ancient Near East, Egypt, Rome, and Greece*, edited by Annette Imhausen and Tanja Pommerening, 277–97. Berlin: De Gruyter.

2016. "Otto Neugebauer's Visits to Copenhagen and His Connection to Denmark." In *A Mathematician's Journeys: Otto Neugebauer and the Modern Transformations of Ancient Science*, edited by Alexander Jones, Christine Proust, and John M. Steele, 107–26. Heidelberg: Springer.

2021. "The Babylonian Dodecatemoria and Calendar Texts: Inverse Schemes for Determining Position and Times for the Schematic Sun and Moon." *Aestimatio* 2, no. 1: 43–64.

2022. "Babylonian House Numbers and Squares of Zodiacal Signs: The Origin of the Quadruplicities, Part I." *Aestimatio* 2, no. 2: 43–58.

Brack-Bernsen, Lis and Hermann Hunger. 2002. "TU 11: A Collection of Rules for the Prediction of Lunar Phases and of Month Lengths." SCIAMVS 3: 3–90.

2008. "BM 42282+42294 and the Goal-Year Method." SCIAMVS 9: 3–23.

Brack-Bernsen, Lis and John M. Steele. 2004. "Babylonian Mathemagics: Two Mathematical Astronomical-Astrological Texts." In *Studies in the History of the Exact Sciences in Honour of David Pingree*, edited by Charles Burnett, Jan P. Hogendijk, Kim Plofker, and Michio Yano, 95–125. Leiden: Brill.

Breger, Gil. 2022. "Ziqpu-Stars and Cuneiform Knowledge: Meaning, Applications, Contexts." PhD diss., University of California, Berkeley.

Brentjes, Sonja. 2010. The Mathematical Sciences in Safavid Iran: Questions and Perspectives." In *Muslim Cultures in the Indo-Iranian World during the Early-Modern and Modern Periods*, edited by Denis Hermann, 325–402. Berlin: F. Speziale.

2018. *Teaching and Learning the Sciences in Islamicate Societies (800–1700)*. Turnhout: Brepols.

Brentjes, Sonja and Jürgen Renn, eds. 2016. *Globalization of Knowledge in the Post-Antique Mediterranean, 700–1500*. London: Routledge.

Bretelle-Establet, Florence, ed. 2010. *Looking at It from Asia: The Processes That Shaped the Sources of History of Science*. Dordrecht: Springer.

Britton, John P. 1987. "The Structure and Parameters of Column φ." In *From Ancient Omens to Statistical Mechanics: Essays on the Exact Sciences Presented to Asger Aaboe*, edited by John Lennart Berggren and Bernard R. Goldstein, 23–26. Copenhagen: University Library.

1989. "An Early Function for Eclipse Magnitudes in Babylonian Astronomy." *Centaurus* 32: 1–52.

1990. "A Tale of Two Cycles: Remarks on Column φ." *Centaurus* 33: 57–69.

1993a. "Scientific Astronomy in Pre-Seleucid Babylon." In *Die Rolle der Astronomie in den Kulturen Mesopotamiens*, edited by Hannes D. Galter, 61–76. Graz: GrazKult.

1993b. "A Table of Fourth Powers and Related Texts from Seleucid Babylon." *Journal of Cuneiform Studies* 43–45: 71–88.

1998. "Planetary Theory in Babylon." *Journal for the History of Astronomy* 29: 381–85.

2002. "Treatments of Annual Phenomena in Cuneiform Sources." In *Under One Sky: Astronomy and Mathematics in the Ancient Near East*, edited by John M. Steele and Annette Imhausen, 21–78. Münster: Ugarit Verlag.

2007a. "Calendars, Intercalations and Year-Lengths in Mesopotamian Astronomy." In *Calendars and Years: Astronomy and Time in the Ancient Near East*, edited by John M. Steele, 115–32. Oxford: Oxbow Books.

2007b. "Studies in Babylonian Lunar Theory: Part I. Empirical Elements for Modeling Lunar and Solar Anomalies." *Archive for History of Exact Sciences* 61: 83–145.

2008. "Remarks on Strassmaier Cambyses 400." In *From the Banks of the Euphrates: Studies in Honor of Alice Slotsky*, edited by Micah Ross, 7–33. Winona Lake, IN: Eisenbrauns.

2009. "Studies in Babylonian Lunar Theory: Part II. Treatments of Lunar Anomaly." *Archive for History of Exact Sciences* 63: 357–431.

2010. "Studies in Babylonian Lunar Theory: Part III. The Introduction of the Uniform Zodiac." *Archive for History of Exact Sciences* 64: 617–63.

Britton, John P., Christine Proust, and Steve Shnider. 2011. "Plimpton 322: A Review and a Different Perspective." *Archive for History of Exact Sciences* 65: 519–66.

Britton, John P. and C. B. F. Walker. 1991. "A Fourth Century Babylonian Model for Venus: BM 33552." *Centaurus* 34: 97–118.

Brown, David. 2000. *Mesopotamian Planetary Astronomy-Astrology*. Leiden: Brill.

Brown, David, Jonathan Ben-Dov, Harry Falk, et al. 2018. *The Interactions of Ancient Astral Science*. Bremen: Hempen.

Brown, David, John Fermor, and Christopher Walker. 1999/2000. "The Water-Clock in Mesopotamia." *Archiv für Orientforschung* 46/47: 130–48.

Brown, Donald E. 1991. *Human Universals*. Philadelphia: Temple University Press.

2004. "Human Universals, Human Nature, and Human Culture." *Daedalus* 133: 47–54.

Burnet, John. 1908. *Early Greek Philosophy*. London: Adam and Charles Black.

Butterfield, Herbert. 1931. *The Whig Interpretation of History*. London: Bell.

Bury, Robert Gregg. 1929. *Plato: Timaeus, Critias, Cleitophon, Menexenus, Epistles*. Cambridge, MA: Harvard University Press.

Candea, Matei. 2010. "For the Motion." *Critique of Anthropology* 30, no. 2: 172–79.

Carrithers, Michael, Matei Candea, Karen Sykes, Martin Holbraad, and Soumhya Venkatesan, eds. 2010. "*Ontology Is Just Another Word for Culture*: Motion Tabled at the 2008 Meeting of the Group for Debates in Anthropological Theory, University of Manchester." *Critique of Anthropology* 30, no. 2: 152–200.

Cassirer, Ernst. (1946) 1953. *Language and Myth*. Translated by Susanne Langer. New York: Harper.

Cat, Jordi. 2021. "The Unity of Science." In *The Stanford Encyclopedia of Philosophy*, edited by Edward N. Zalta. Fall 2021. https://plato.stanford.edu/archives/fall2021/entries/scientific-unity.

Chang, Hasok. 2012. *Is Water H2O?: Evidence, Realism and Pluralism*. New York: Springer.

2022. *Realism for Realistic People: A New Pragmatist Philosophy of Science*. Cambridge: Cambridge University Press.

Chemla, Karine. 2012. *The History of Mathematical Proof in Ancient Traditions*. Cambridge: Cambridge University Press.

Chicago Assyrian Dictionary. 1956–2010. Vols. 1–21. Chicago: University of Chicago, Oriental Institute.

Chumley, Lily. 2017. "Qualia and Ontology: Language, Semiotics, and Materiality; An Introduction." *Signs and Society* 5, no. S1: S1–S20.

Civil, Miguel. 1969. *The Series lú = ša and Related Texts: Materials for the Sumerian Lexicon (MSL), XII*. With the collaboration of R. D. Biggs, Hans

G. Güterbock, Hans J. Nissen, and E. Reiner. Rome: Pontificium Institutum Biblicum.
Cladis, Mark S. 1991. "Durkheim, Goodman, Rorty, and Mild-Mannered Pragmatism: French and American Style." *Stanford French Review* 15: 139–63.
Collingwood, R. G. 1945. *The Idea of Nature*. Oxford: Clarendon Press.
Collins, Harry. 2008. "Actors' and Analysts' Categories in the Social Analysis of Science." In *Clashes of Knowledge. Knowledge and Space*, vol. 1, edited by Peter Meusburger, Michael Welker, and Edgar Wunder, 101–10. Dordrecht: Springer.
Conant, James and John Haugeland, eds. 2000. *The Road since Structure: Thomas S. Kuhn. Philosophical Essays, 1970–1993, with an Autobiographical Interview*. Chicago: University of Chicago Press.
Cooke, Roger L. 1997. *The History of Mathematics: A Brief Course*. New York: Wiley.
Copenhaver, Brian. 1988. "Natural Philosophy: Astrology and Magic." In *The Cambridge History of Renaissance Philosophy*, edited by Charles B. Schmitt, Quentin Skinner, Eckhard Kessler, and Jill Kraye, 264–300. Cambridge: Cambridge University Press.
Crisostomo, Jay. 2018. "Language, Translation, and Commentary in Cuneiform Scribal Practice." *Journal of Ancient Near Eastern History* 5: 41–56.
2019. *Translation as Scholarship: Language, Writing, and Bilingual Education in Ancient Babylonia*. Berlin: De Gruyter.
Cruse, D. A. 1986. *Lexical Semantics*. Cambridge: Cambridge University Press.
Cumont, Franz, Franz Boll, and Wilhelm Kroll, eds. 1898–1953. *Catalogus Codicum Astrologorum Graecorum*. 12 vols. Brussels: Lamertin.
Daston, Lorraine. 2016. "History of Science without 'Structure.'" In *Kuhn's "Structure of Scientific Revolutions" at Fifty: Reflections on a Science Classic*, edited by Robert J. Richards and Lorraine Daston, 115–32. Chicago: University of Chicago Press.
Davidson, Donald. 1973. "On the Very Idea of a Conceptual Scheme." *Proceedings and Addresses by the American Philosophical Association* 43: 5–20.
1984. *Inquiries into Truth and Interpretation*. Oxford: Oxford University Press.
2005. *Truth, Language and History*. Oxford: Oxford University Press.
Dear, Peter. 2005. "What Is the History of Science the History Of? Early Modern Roots of the Ideology of Modern Science." *Isis* 96: 390–406.
2012. "Science Is Dead: Long Live Science." *Osiris* 27: 37–55.
Declos, Alexandre. 2019. "Goodman's Many Worlds." *Journal for the History of Analytical Philosophy* 7: 1–25.
De Jong, Teije. 2016. "The Players and the Field." In *A Mathematician's Journeys: Otto Neugebauer and the Modern Transformations of Ancient Science*, edited by Alexander Jones, Christine Proust, and John M. Steele, 265–302. Heidelberg: Springer.
2019. "A Study of Babylonian Planetary Theory I: The Outer Planets." *Archive for History of Exact Sciences* 73: 1–37.

De Zorzi, Nicla. 2014. *La Serie Teratomantica Šumma Izbu: Testo, Tradizione, Orizzonti Culturali*, 2 vols. Padua: S.A.R.G.O.N. Editrice e Libreria.

2017. "Teratomancy at Tiguananum: Structure, Hermeneutics and 'Weltanschauung' of a Northern Mesopotamian Omen Corpus." *Journal of Cuneiform Studies* 69: 125–50.

Delaporte, François, ed. 2000. *A Vital Rationalist: Selected Writings from Georges Canguilhem*. Translated by Arthur Goldhammer. New York: Zone Books.

Delnero, Paul. 2010. "Sumerian Literary Catalogues and the Scribal Curriculum." *Zeitschrift für Assyriologie* 100: 32–55.

Descola, Philippe. [1986] 1994. *In the Society of Nature: A Native Ecology in Amazonia*. Translated by Nora Scott. Cambridge: Cambridge University Press.

2010. "Cognition, Perception and Worlding." *Interdisciplinary Science Reviews* 35: 334–40.

2013. *Beyond Nature and Culture*. Translated by Janet Lloyd. Chicago: University of Chicago Press.

2014. "Modes of Being and Forms of Predication." *HAU: Journal of Ethnographic Theory* 4: 271–80.

Descola, Philippe and Gísli Pálsso, eds. 1996. *Nature and Society: Anthropological Perspectives*. London: Routledge.

Diakonoff, Igor M. 1961. "More on Assyriology." *Current Anthropology* 2: 390–91.

Drawnel, Henryk. 2011. *The Aramaic Astronomical Book (4Q208–4Q211) from Qumran: Text, Translation, and Commentary*. Oxford: Oxford University Press.

Drower, Ethel Stefana. 1949. *The Book of the Zodiac (Sfar Malwasia)*. London: Brill.

Dutra, Bruce P. 1999. "The Problem of Conflicting Truths and Nelson Goodman's Ontological Pluralism." PhD diss., Purdue University.

Elgin, Catherine Z. 1984. "Goodman's Rigorous Relativism." *Journal of Thought* 19: 36–45.

2000. "Worldmaker: Nelson Goodman." *Journal for General Philosophy of Science* 31: 1–18.

Ellis, Brian David. 1985. "What Science Aims to Do." In *Images of Science*, edited by Paul M. Churchland and Clifford A. Hooker, 48–74. Chicago: University of Chicago Press.

1988. "Internal Realism." *Synthese* 76: 409–34.

Ellison, Ralph. 1952. *Invisible Man*. New York: Random House.

Epping, Joseph, and Johann Strassmaier. 1889. *Astronomisches aus Babylon, oder das Wissen der Chaldäer über den gestirnten Himmel: Stimmen aus Maria-Laach*. Freiburg im Breisgau: Herder'sche Verlagshandlung.

Escobar, Eduardo A. 2017. "Technology as Knowledge: Cuneiform Technical Recipes and the Material World." PhD diss., University of California, Berkeley.

2018. "Bricoleurs in Babylonia: The Scribes of Enūma Anu Enlil." In *"The Scaffolding of Our Thoughts": Essays on Assyriology and the History of*

Science in Honor of Francesca Rochberg, edited by C. Jay Crisostomo, Eduardo A. Escobar, Terri Tanaka, and Niek Veldhuis, 264–85. Leiden: Brill.

2019a. "Glassmaking as Scribal Craft." In *A Wonder to Behold: Craftsmanship and the Creation of Babylon's Ishtar Gate*, edited by Anastasia Amrhein, Clare Fitzgerald, and Elizabeth Knott, 119–125. Princeton, NJ: Princeton University Press.

2019b. "Number Culture: Old Babylonian Mathematics at Yale." In *Ancient Mesopotamia Speaks: Highlights of the Yale Babylonian Collection*, edited by Agnete W. Lassen, Eckart Frahm, and Klaus Wagensonner, 185–91. New Haven, CT: Yale University Press.

2022. "Colour and Cultural Maintenance in Ancient Babylonia: Contextualizing Wool Dye Recipe BM 62788 + BM 82978." *Technai* 13: 67–84.

2023. "Tappūtī-Bēlet-Ekallim (fl. 1200 BCE): A Cuneiform Tablet on Middle Assyrian Perfumery (c. 1200 BCE)." In *Women in the History of Science: A Sourcebook*, edited by Hannah Wills, Sadie Harrison, Erika Jones, Farrah Lawrence-Mackey, and Rebecca Martin, 15–22. London: UCL Press.

ETCSL project. 2006. "Enki and the World Order." http://etcsl.orinst.ox.ac.uk/cgi-bin/etcsl.cgi?text=t.1.1.3#.

Evans, James C. 2020. "The Ptolemaic Planetary Hypotheses." In *Hellenistic Astronomy: The Science in Its Contexts*, edited by Alan C. Bowen and Francesca Rochberg, 112–24. Leiden: Brill.

Evans, James C., and J. Lennart Berggren. 2006. *Geminos's Introduction to the Phenomena: A Translation and Study of a Hellenistic Survey of Astronomy*. Princeton, NJ: Princeton University Press.

Fincke, Jeanette C. 2004. "The British Museum's Ashurbanipal Library Project." *Iraq* 66: 55–60.

Finkel, Irving L. 1988. "Adad-apla-iddina, Esagil-kīn-apli and the Series SA.GIG." In *A Scientific Humanist: Studies in Memory of Abraham Sachs*, edited by Erle Liechty, Maria de J. Ellis, and Pamela Gerardi, 143–59. Philadelphia: University Museum.

2000. "On Late Babylonian Medical Training." In *Wisdom, God and Literature: Studies in Assyriology in Honor of W. G. Lambert*, edited by Andrew R. George and Irving L. Finkel, 137–224. Winona Lake, IN: Eisenbrauns.

2019. "Assurbanipal's Library: An Overview." In *Libraries before Alexandria: Ancient Near Eastern Traditions*, edited by Kim Ryholt and Gojko Barjamovic, 367–89. Oxford: Oxford University Press.

Finkel, Irving L., and Markham J. Geller, eds. 2007. *Disease in Babylonia*. Leiden: Brill.

Finkel, Irving L., and Julian E. Reade. 1996. "Assyrian Hieroglyphs." *Zeitschrift für Assyriologie* 86: 244–68.

Finkel, Irving L., and Michael J. Seymour, eds. 2009. *Babylon*. Oxford: Oxford University Press.

Finkelstein, J. J. 1962. "Mesopotamia." *Journal of Near Eastern Studies* 21: 73–92.

Fleck, Ludwik. 1935. *Entstehung und Entwicklung einer wissenschaftlichen Tatsache: Einführung in die Lehre vom Denkstil und Denkkollektiv*. Edited and with a foreword by Lothar Schäfer and Thomas Schnelle. Basel: Benno Schwabe.

1979. *Genesis and Development of a Scientific Fact*, edited by Thaddeus J. Trenn and Robert K. Merton. Translated by Fred Bradley and Thaddeus J. Trenn. Foreword by Thomas S. Kuhn. Chicago: University of Chicago Press.

Forster, E. M. 1923. *Pharos and Pharillion*. New York: Knopf.

Frahm, Eckart. 2005. "A Note on Nabû-zuqup-kenu." *N.A.B.U.* 1.

2011a. *Babylonian and Assyrian Text Commentaries: Origins of Interpretation*. Münster: Ugarit Verlag.

2011b. "Keeping Company with Men of Learning: The King as Scholar." In *The Oxford Handbook of Cuneiform Culture*, edited by Karen Radner and Eleanor Robson, 508–32. Oxford: Oxford University Press.

Frede, Michael, and David Charles. 2001. *Aristotle Metaphysics Lambda*. Oxford: Clarendon Press.

Friberg, Jöran. 2007. *A Remarkable Collection of Babylonian Mathematical Texts*. Berlin: Springer.

2008. "A Remarkable Collection of Babylonian Mathematical Texts." *Notices of the American Mathematical Society* 55: 1076–86.

Geertz, Clifford. (1973) 2000. *Interpretation of Cultures*. New York: Basic Books.

(1983) 2000. *Local Knowledge: Further Essays in Interpretive Anthropology*. 2nd ed. New York: Basic Books.

1984. "Distinguished Lecture: Anti-Anti-Relativism." *American Anthropologist* 86: 263–78.

1986. "The Uses of Diversity." *Michigan Quarterly Review* 25, no. 1: 105–23. Also printed in *The Tanner Lectures on Human Values* 7: 253–75.

2000. *Available Light*. Princeton, NJ: Princeton University Press.

2005. "Deep Play: Notes on the Balinese Cockfight." *Daedalus* 134: 56–86.

2010. *Life among the Anthros, and Other Essays*, edited by Fred Inglis. Princeton, NJ: Princeton University Press.

Gehlken, Erlend. 2005. "Die Adad-Tafeln der Omenserie Enuma Anu Enlil. Teil 1: Einführung." *Baghdader Mitteilungen* 36: 235–73.

Geller, Markham J. 1999–2000. "The Landscape of the Netherworld." In *Landscapes: Territories, Frontiers and Horizons in the Ancient Near East. Papers Presented to the XLIV Rencontre Assyriologique Internationale, Venezia, 7–11 July 1997*, edited by L. Milano, S. de Martino, F. M. Fales, and G. B. Lanfranchi, 41–49. Padua: Sargon.

2000. Review of Alice Slotsky, *The Bourse of Babylon: Market Quotations in the Astronomical Diaries of Babylonia*, 1997. *Orientalische Literaturzeichnung* 95: 410–12.

2010. *Ancient Babylonian Medicine: Theory and Practice*. Oxford: Wiley-Blackwell.

2018. "The Exorcist's Manual (KAR 44)." In *Assyrian and Babylonian Scholarly Text Catalogues: Medicine, Magic and Divination*, edited by Ulrike Steinert, 292–304. Boston: De Gruyter.

2020. "An Apothecary's Handbook." *Le journal des medecines cuneiformes* 35: 1–33.
2022. "Second Order Lexicography." *Archiv für Orientforschung* 55: 87–95.
Gellner, Ernest. 1974. *Legitimation of Belief*. London: Cambridge University Press.
1979. *Spectacles and Predicaments: Essays in Social Theory*. Cambridge: Cambridge University Press.
1982. "Relativism and Universals." In *Rationality and Relativism*, edited by Martin Hollis and Steven Lukes, 188–200. Cambridge, MA: MIT Press.
1985. *Relativism and the Social Sciences*. Cambridge: Cambridge University Press.
George, Andrew R. 2010. "The Epic of Gilgamesh." In *The Cambridge Companion the the Epic*, edited by Catherine Bates, 1–12. Cambridge: Cambridge University Press.
George, Andrew R., and Farouk N. H. Al-Rawi. 1991. "Enūma Anu Enlil XIV and Other Early Astronomical Tables." *Archiv für Orientforschung* 38/39: 52–73.
Geminus. 2006. *Geminos' Introduction to the Phenomena: A Translation and Study of a Hellenistic Survey of Astronomy*. Edited and translated by James Evans and J. Lennart Berggren. Princeton, NJ: Princeton University Press.
George, Andrew. 1986. Sennacherib and the Tablet of Destinies. *Iraq* 48: 133–46.
1992. *Babylonian Topographical Texts*. Leuven: Peeters.
1999a. *The Epic of Gilgamesh: The Babylonian Epic Poem and Other Akkadian and Sumerian Texts*. London: Penguin Books.
1999b. "E-SANGIL and E-TEMEN-ANKI, the Archetypal Cult-Centre." In *Babylon: Focus mesopotamischer Geschichte, Wiege früher Gelehrsamkeit, Mythos in der Moderne*, edited by Johannes Renger, 1–14. Saarbrücken: SDV Saarbrücken.
Getachew, Adom. 2019. *Worldmaking after the Rise and Fall of Self-Determination*. Princeton, NJ: Princeton University Press.
Glock, Hans-Johann. 2003. *Quine and Davidson on Language, Thought and Reality*. Cambridge: Cambridge University Press.
Goetze, Albrecht. 1959. "The 38th Tablet of the Series á-A-naqu." *Journal of Cuneiform Studies* 13: 120–27.
Goldstein, Bernard R. 1967. "The Arabic Version of Ptolemy's Planetary Hypotheses." *Transactions of the American Philosophical Society* 57: 3–55.
1980. "The Status of Models in Ancient and Medieval Astronomy." *Centaurus* 24: 132–47.
1997. "Saving the Phenomena: The Background to Ptolemy's Planetary Theory." *Journal for the History of Astronomy* 28: 1–12.
2003. "Ancient and Medieval Values for the Mean Synodic Month." *Journal for the History of Astronomy* 34: 65–74.
Goldstein, C. 2000. "Egypt and Mesopotamia." In *Reader's Guide to the History of Science*, edited by Arne Hessenbruch, 195–97. London: Fitzroy Dearborn.
Golinski, Jan. 2012. "Is It Time to Forget Science? Reflections on Singular Science and Its History." *Osiris* 27, no. 1: 19–36.

Gomperz, Theodor. (1893) 1911. *Griechische Denker: Eine Geschichte der antiken Philosophie*. 3 vols. 3rd ed. Berlin: De Gruyter.

Goodman, Nelson. (1951) 1977. *The Structure of Appearance*. 3rd ed. Dordrecht: Reidel.

(1954) 1983. *Fact, Fiction, and Forecast*. 4th ed. Cambridge, MA: Harvard University Press.

1960. "The Way the World Is." *The Review of Metaphysics* 14: 48–56. Republished in Peter J. McCormick, ed., *Starmaking: Realism, Anti-Realism, and Irrealism*. Cambridge, MA: MIT Press (1996), 3–10.

1968. *Languages of Art*. Indianapolis, IN: Hackett.

Goodman, Nelson, ed. 1972. *Problems and Projects*. Indianapolis, IN: Bobbs Merrill.

Goodman, Nelson. 1975. "Words, Works, Worlds." *Erkenntnis* 9: 57–73.

1978. *Ways of Worldmaking*. Indianapolis, IN: Hackett.

1983. "Notes on the Well-Made World." *Erkenntnis* 19: 99–107.

1984. *Of Mind and Other Matters*. Cambridge, MA: Harvard University Press.

(1989) 2010. "Just the Facts, Ma'am!" In *Relativism: Interpretation and Confrontation*, edited by Michael Krausz, 80–85. Notre Dame, IN: University of Notre Dame Press. Republished in Michael Krausz, ed., Relativism: A Contemporary Anthology. New York: Columbia University Press (2010), pp. 80–85.

1993. "On Some Worldly Worries." *Synthese* 96: 9–12.

1996a. "On Starmaking." In *Starmaking: Realism, Anti-Realism and Irrealism*, edited by Peter J. McCormick, 143–47. Cambridge, MA: MIT Press.

1996b. "The Way the World Is." In *Starmaking: Realism, Anti-Realism and Irrealism*, edited by Peter J. McCormick, 3–10. Cambridge, MA: MIT Press.

Goodman, Nelson, and Catherine Z. Elgin. 1988. *Reconceptions in Philosophy and Other Arts and Sciences*. Indianapolis, IN: Hackett.

Goodman, Nelson, and Willard von Orman Quine. 1947. "Steps toward a Constructive Nominalism." *The Journal of Symbolic Logic* 12: 105–22.

Goody, Jack. (1977) 1995. *Domestication of the Savage Mind*. Cambridge: Cambridge University Press.

Gosselin, Mia. 1990. *Nominalism and Contemporary Nominalism: Ontological and Epistemological Implications of the Work of W. V. O. Quine and of N. Goodman*. Dordrecht: Kluwer Academic.

Graeber, David. 2015. "Radical Alterity Is Just Another Way of Saying 'Reality': A Reply to Eduardo Viveiros de Castro." *HAU: Journal of Ethnographic Theory* 5: 1–41.

Grafton, Anthony. 1999. *Cardano's Cosmos*. Cambridge, MA: Harvard University Press.

Gray, J. M. K., and John M. Steele. 2008. "Studies on Babylonian Goal-Year Astronomy I: A Comparison between Planetary Data in Goal-Year Texts, Almanacs and Normal Star Almanacs." *Archive for History of Exact Sciences* 62: 553–600.

2009. "Studies on Babylonian Goal-Year Astronomy II: The Babylonian Calendar and Goal-Year Methods of Prediction." *Archive for History of Exact Sciences* 63: 611–33.

Greenfield, Jonas, and Alan Sokoloff. 1989. "Astrological and Related Omen Texts in Jewish Palestinian Aramaic." *Journal of Near Eastern Studies* 48: 201–14.

Gumplowicz, Ludwig. 1879. *Das Recht der Nationalität und Sprachen in Oesterreich-Ungarn*. Innsbruck: Wagner'sche Universitäts Buchhandlung.

Güterbock, H. G. 1987. "Hittite Liver Models." In *Language, Literature, and History: Philological and Historical Studies Presented to Erica Reiner*, edited by Francesca Rochberg-Halton, 147–53. New Haven, CT: American Oriental Society.

1988. "Bilingual Moon Omens from Boghazköy." In *A Scientific Humanist: Studies in Memory of Abraham Sachs*, edited by Erle Leichty, Maria de J. Ellis, and Pamela Gerardi, 161–74. Philadelphia: University Museum.

Guthrie, W. K. C. 1971. *Aristotle. VI. On the Heavens*. Cambridge, MA and London: Harvard University Press and William Heinemman Ltd.

Hacking, Ian. 1988. "The Participant Irrealist at Large in the Laboratory." *The British Journal for the Philosophy of Science* 39: 277–94.

1999. *The Social Construction of What?* Cambridge, MA: Harvard University Press.

2002. *Historical Ontology*. Cambridge, MA: Harvard University Press.

Hallo, William H. 2010. *The World's Oldest Literature: Studies in Sumerian Belles-Lettres*. Leiden: Brill.

Hanegraaff, Wouter J. 2012. *Esotericism and the Academy: Rejected Knowledge in Western Culture*. Cambridge: Cambridge University Press.

Hanks, William F. and Carlo Severi. 2014. "Translating Worlds: The Epistemological Space of Translation." *HAU: Journal of Ethnographic Theory* 4, no. 14: 1–16.

Hanson, Norwood Russell. 1958. *Patterns of Discovery: An Inquiry into the Conceptual Foundations of Science*. Cambridge: Cambridge University Press.

Hansson, Sven Ove. 2009. "Cutting the Gordian Knot of Demarcation." *International Studies in the Philosophy of Science* 23: 237–43.

2013. "Defining Pseudoscience and Science." In *Philosophy of Pseudoscience: Reconsidering the Demarcation Problem*, edited by Massimo Pigliucci and Maarten Boudry, 61–78. Chicago: University of Chicago Press.

Haraway, Donna. 1988. "Situated Knowledges: The Science Question in Feminism as a Site of Discourse on the Privilege of Partial Perspective." *Feminist Studies* 14: 575–99.

Harman, Oren, and Alexandre Métraux, eds. 2013. "Approaches, Styles and Narratives: Reflections on Doing History of Science." *Science in Context* 26: 553–664.

Harrison, Peter. 2006. "Disjoining Wisdom and Knowledge: Science, Theology, and the Making of Western Modernity." In *Wisdom or Knowledge? Science, Theology, and Cultural Dynamics*, edited by Willem B. Drees, Hubert Meisinger, and Zbigniew Liana, 51–72. London: T&T Clark.

2017. "Science and Secularization." *Intellectual History Review* 27: 47–70.

2019. "Introduction." In *Science without God? Rethinking the History of Scientific Naturalism*, edited by Peter Harrison and Jon H. Roberts, 1–18. Oxford: Oxford University Press.

Harrison, Peter, and Jon H. Roberts, eds. 2019. *Science without God? Rethinking the History of Scientific Naturalism*. Oxford: Oxford University Press.

Haubold, Johannes, John Steele, and Kathryn Stevens. 2019. *Keeping Watch in Babylon: The Astronomical Diaries in Context*. Leiden: Brill.

Heeßel, Nils P. 2000. *Babylonisch-Assyrische Diagnostik*. Münster: Ugarit Verlag.

2005. "Stein, Pflanze und Holz: Ein neuer Text zur 'medizinischen Astrologie.'" *Orientalia* 74: 1–22.

2008. "Astrological Medicine in Babylonia." In *Astro-Medicine, Astrology and Medicine, East and West*, edited by Anna Akasoy, Charles Burnett, and Ronit Yoeli-Tlalim, 1–16. Florence: Sismel–Edizioni del Galluzzo.

2021. *Divinatorische Texte III: Astrologische Omina*. Wiesbaden: Harrassowitz Verlag.

Heimpel, Wolfgang. 1986. "The Sun at Night and the Doors of Heaven in Babylonian Texts." *Journal of Cuneiform Studies* 38: 127–51.

Hellman, Geoffrey. 1977. Introduction to Nelson Goodman, *The Structure of Appearance*, 3rd ed., xix–xlvii. Dordrecht: D. Reidel Publishing Company.

Henare, Amiria, Martin Holbraad, and Sari Wastell, eds. 2007. *Thinking through Things: Theorizing Artefacts Ethnographically*. London: Routledge.

Henning, W. B. 1942. "An Astronomical Chapter of the *Bundahishn*." *Journal of the Royal Asiatic Society* 74: 229–48.

Hessenbruch, Arne. 2000. *Reader's Guide to the History of Science*. London: Routledge.

Heywood, Paolo. 2012. "Anthropology and What There Is: Reflections on 'Ontology.'" *The Cambridge Journal of Anthropology* 30: 143–51.

2022. "The Ontological Turn." In *The Cambridge Encyclopedia of Anthropology*. https://www.anthroencyclopedia.com/entry/ontological-turn.

Hiley, David R., James Bohman, and Richard Shusterman. 1991. *The Interpretive Turn: Philosophy, Science, Culture*. Ithaca, NY: Cornell University Press.

Hoffman, Joshua, and Gary S. Rosenkrantz. 1997. *Substance: Its Nature and Existence*. London: Routledge.

Holbraad, Martin. 2012. *Truth in Motion: The Recursive Anthropology of Cuban Divination*. Chicago: University of Chicago Press.

Holbraad, Martin, and Morten Axel Pedersen. 2017. *The Ontological Turn: An Anthropological Exposition*. Cambridge: Cambridge University Press.

Holmes, Brooke. 2019. "On Stoic Sympathy: Cosmobiology and the Life of Nature." In *Antiquities beyond Humanism*, edited by Emanuela Bianchi, Sara Brill, and Brooke Holmes, 239–70. Oxford: Oxford University Press.

Horowitz, Wayne. 1994. "Two New Ziqpu-Star Texts and Stellar Circles." *Journal of Cuneiform Studies* 46: 89–98.

(1998) 2017. *Mesopotamian Cosmic Geography*. Winona Lake, IN: Eisenbrauns.

2000. "Astral Tablets in the Hermitage Saint Petersburg." *Zeitschrift für Assyriologie* 90: 194–206.

2014. *The Three Stars Each: The Astrolabes and Related Texts*. Vienna: Institut für Orientalistik der Universität Wien.

Horton, Robin. 1967a. "African Traditional Thought and Western Science. Part I: From Tradition to Science." *Africa: Journal of the International African Institute* 37: 50–71.

1967b. "African Traditional Thought and Western Science. Part II: The 'Closed' and 'Open' Predicaments." *Africa: Journal of the International African Institute* 37: 155–87.

1982. "Tradition and Modernity Revisited." *In Rationality and Relativism*, edited by Martin Hollis and Steven Lukes, 201–60. Cambridge, MA: MIT Press.

1997. *Patterns of Thought in Africa and the West: Essays on Magic, Religion and Science*. Cambridge: Cambridge University Press.

Hoskin, Michael. 1998. "Olaf Pedersen 1920–1998." *Astronomy and Geophysics* 39, no. 5: 5.33–5.34.

Hoyningen-Huene, Paul. 1990. "Kuhn's Conception of Incommensurability." *Studies in History and Philosophy of Science* 21: 481–92.

1993. *Reconstructing Scientific Revolutions*. Chicago: University of Chicago Press.

Hoyrup, Jens. 1990. "Algebra and Naive Geometry: An Investigation of Some Basic Aspects of Old Babylonian Mathematical Thought I." *Altorientalische Forschungen* 17: 27–69.

2017. *Algebra in Cuneiform: Introduction to an Old Babylonian Geometrical Technique*. Berlin: Edition Open Access, Max Planck Institute for the History of Science.

Hsia, Florence C. 2009. *Sojourners in a Strange Land: Jesuits and Their Scientific Missions in Late Imperial China*. Chicago: University of Chicago Press.

2017. "Astronomy after the Deluge." In *Science in the Archives: Pasts, Presents, Futures*, edited by Lorraine Daston, 17–52. Chicago: University of Chicago Press.

Hughes, R. I. G. 1997. "Models and Representation." *Philosophy of Science* 64, no. S4: S325–36.

Hunger, Hermann. 1968. *Babylonische und Assyrische Kolophone*. Neukirchen-Vluyn: Verlag Butzon & Bercker Kevelaer.

1976. "Astrologische Wettervorhersagen." *Zeitschrift für Assyriologie und Vorderasiatische Archäologie* 66: 234–60.

1992. *Astrological Reports to Assyrian Kings*. Helsinki: Helsinki University Press.

1999. "Non-Mathematical Astronomical Texts and Their Relationships." In *Ancient Astronomy and Celestial Divination*, edited by Noel M. Swerdlow, 77–96. Cambridge, MA: MIT Press.

2001. *Astronomical Diaries and Related Texts from Babylonia. Vol. 5: Lunar and Planetary Texts*. Vienna: Österreichische Akkademie der Wissenschaften.

2006. *Astronomical Diaries and Related Texts from Babylonia. Vol. 6: Goal Year Texts*. Vienna: Österreichische Akkademie der Wissenschaften.

2022. *Astronomical Diaries and Related Texts from Babylonia. Vol. 4: Undated Diaries and Addenda, Including Materials by A. J. Sachs*.Vienna: Österreichische Akademie der Wissenschaften.

Hunger, Hermann, and Teije de Jong. 2014. "Almanac W22340A from Uruk: The Last Datable Cuneiform Tablet." *Zeitschrift für Assyriologie und Vorderasiatische Archäologie* 104: 182–94.

Hunger, Hermann, and David Pingree. 1999. *Astral Sciences in Mesopotamia*. Leiden: Brill.

Hunger, Hermann, and Abraham J. Sachs. 2001. *Astronomical Diaries and Related Texts from Babylonia. Vol. 5: Lunar and Planetary Texts*. Vienna: Verlag der Österreichischen Akademie der Wissenschaften, Wien.

Hunger, Hermann, and John Steele. 2019. *The Babylonian Astronomical Compendium MUL.APIN*. New York: Routledge.

Hyspicles. *Anaphorikos*. 1966. Edited and translated by Vittorio de Falco and Max Krause with Otto Neugebauer. Göttingen: Vandenhoeck & Ruprecht.

Ierodiakonou, Katerina. 2006. "Stoic Logic." In *A Companion to Ancient Philosophy*, edited by Mary Louise Gill and Pierre Pellegrin, 505–29. Oxford: Blackwell.

Imhausen, Annette, and Tanja Pommerening, eds. 2010. *Writings of Early Scholars in the Ancient Near East, Egypt, Rome, and Greece*. Berlin: De Gruyter.

2016. *Translating Writings of Early Scholars in the Ancient Near East, Egypt, Greece and Rome. Methodological Aspects with Examples*. Berlin: De Gruyter.

Inglis, David, and John Bone. 2006. "Boundary Maintenance, Border Crossing and the Nature/Culture Divide." *European Journal of Social Theory* 9: 272–87.

Ingold, Tim. 2000. *The Perception of the Environment: Essays on Livelihood, Dwelling and Skill*. London: Routledge.

Jacobus, Helen. 2014. "Greco-Roman Zodiac Sundials and Their Links to a Qumran Calendar (4Q208–4Q209)." *Mediterranean Archaeology and Archaeometry* 14: 67–81.

Jardine, Nicholas. 1986. *The Fortunes of Inquiry*. Oxford: Clarendon Press.

2000. "Uses and Abuses of Anachronism in the History of Science." *History of Science* 38: 251–70.

2019. "Turning to Ontology in Studies of Distant Sciences." *HAU: Journal of Ethnographic Theory* 9: 172–78.

Jensen, Peter. 1890. *Die Kosmologie der Babylonier*. Strasbourg: K. J. Trübner.

Johnson, J. Cale, and Alessandro Stavru, eds. 2019. *Visualizing the Invisible with the Human Body: Physiognomy and Ekphrasis in the Ancient World*. Berlin: Walter de Gruyter.

Jones, Alexander. 1991. "The Adaptation of Babylonian Methods in Greek Numerical Astronomy." *Isis* 82: 440–53.

1996. "On Babylonian Astronomy and Its Greek Metamorphoses." In *Tradition, Transmission, and Transformation*, edited by F. Jamil Rajep and Sally P. Rajep, with Steven Livesey, 139–55. Leiden: Brill.

1999. *Astronomical Papyri from Oxyrhynchus*. Vols. 1–2. Philadelphia: American Philosophical Society.

2016. "Translating Greek Astronomy: Theon of Smyrna on the Apparent Motions of the Planets." In *Translating Writings of Early Scholars in the*

Ancient Near East, Egypt, Greece and Rome: Methodological Aspects with Examples, edited by Annette Imhausen and Tanja Pommerening, 465–505. Berlin: De Gruyter.

Jones, Alexander, Christine Proust, and John M. Steele, eds. 2016. *A Mathematician's Journeys: Otto Neugebauer and the Modern Transformations of Ancient Science*. Heidelberg: Springer.

Jones, Alexander and John M. Steele. 2011. "A New Discovery of a Component of Greek Astrology in Babylonian Tablets: The '*Terms*.'" *ISAW Papers* 1. http://dlib.nyu.edu/awdl/isaw/isaw-papers/1.

2018. "Diodorus on the Chaldeans." In *"The Scaffolding of Our Thoughts": Essays on Assyriology and the History of Science in Honor of Francesca Rochberg*, edited by C. Jay Crisostomo, Eduardo A. Escobar, Terri Tanaka, and Niek Veldhuis, 333–52. Leiden: Brill.

Jordheim, Helge and David Gary Shaw. 2020. "History of Knowledge." Special issue, *History and Theory* 59, no. 4.

Katz, Dina. 2003. *The Image of the Netherworld in the Sumerian Sources*. Bethesda, MD: CDL Press.

Kelley, Donald R. (2002) 2017. *The Descent of Ideas: The History of Intellectual History*. Farnham: Ashgate.

King, L. W. 1901. *Cuneiform Texts from Babylonian Tablets in the British Museum, Vol. 13*. London: The Trustees of the British Museum.

1902. *The Seven Tablets of Creation, or The Babylonian and Assyrian Legends concerning the Creation of the World and of Mankind*. London: Luzac.

Kirk, G. S., J. E. Raven, and M. Schofield. 1983. *The Presocratic Philosophers: A Critical History with a Selection of Texts*. 2nd ed. Cambridge: Cambridge University Press.

Koch, Johannes. 1989. *Neue Untersuchungen zur Topographie des babylonischen Fixternhimmels*. Wiesbaden: Harrassowitz Verlag.

Koch, Ulla Susanne. 2005. *Secrets of Extispicy: The Chapter Multābiltu of the Babylonian Extispicy Series and Niṣirti Bārûti Texts Mainly from Aššurbanipal's Library*. Münster: Ugarit Verlag.

Koch-Westenholz, Ulla. 1993. "Mesopotamian Astrology at Hattusas." In *Die Rolle der Astronomie in den Kulturen Mesopotamiens*, edited by Hannes D. Galter, 231–46. Graz: GrazKult.

1995. *Mesopotamian Astrology: An Introduction to Babylonian and Assyrian Celestial Divination*. Copenhagen: Museum Tusculanum Press.

2000. *Babylonian Liver Omens: The Chapters Manzazu, Padanu and Pan Takalti of the Babylonian Extispicy Series Mainly from Assurbanipals's Library*. Copenhagen: Museum Tusculanum Press.

Köcher, Franz. 1963–. *Die babylonisch-assyrische Medizin in Texten und Untersuchungen*. Berlin: De Gruyter.

Kohn, Eduardo. 2014. "How Dogs Dream: Amazonian Natures and the Politics of Transspecies Engagement." *American Ethnologist* 34: 3–24.

Kondo, Dorinne. 2018. *Worldmaking: Race, Performance, and the Work of Creativity*. Durham, NC: Duke University Press.

Koyré, Alexandre. 1957. *From the Closed World to the Infinite Universe*. Baltimore, MD: Johns Hopkins University Press.

Kugler, F. X. 1900. *Die Babylonische Mondrechnung: Zwei Systeme der Chaldäer über den Lauf des Mondes und der Sonne*. Freiburg im Breisgau: Herder'sche Verlagshandlung.

Kuhn, Thomas S. 1957. *The Copernican Revolution: Planetary Astronomy in the Development of Western Thought*. Cambridge, MA: Harvard University Press.

(1962) 1996. *The Structure of Scientific Revolutions*. 3rd ed. Chicago: University of Chicago Press.

(1989) 2000. "Possible Worlds in History of Science." In *Possible Worlds in Humanities, Arts and Sciences: Proceedings of Nobel Symposium 65*, edited by Sture Allén, 9–32. Berlin: De Gruyter. Republished in James Conant and John Haugeland, eds., *The Road since Structure: Thomas S. Kuhn. Philosophical Essays, 1970–1993, with an Autobiographical Interview*. Chicago: University of Chicago Press, 2000, pp. 58–89.

(1991) 2000a. "The Road Since Structure." *Proceedings of the Biennial Meeting of the Philosophy of Science Association* 2: 3–13. Republished in James Conant and John Haugeland, eds., *The Road since Structure: Thomas S. Kuhn. Philosophical Essays, 1970–1993, with an Autobiographical Interview*. Chicago: University of Chicago Press, 2000, pp. 90–104.

(1991) 2000b. "The Natural and the Human Sciences." In *The Interpretive Turn: Philosophy, Science, Culture*, edited by David Hiley, James Bohman, and Richard Shusterman, 17–24. Ithaca, NY: Cornell University Press. Republished in James Conant and John Haugeland, eds., *The Road since Structure: Thomas S. Kuhn. Philosophical Essays, 1970–1993, with an Autobiographical Interview*. Chicago: University of Chicago Press, 2000, pp. 216–23.

Labat, René. 1988. *Manuel d'epigraphie akkadienne*. 6th ed. Paris: Paul Geuthner.

Ladyman, James. 2013. "Toward a Demarcation of Science from Pseudoscience." In *Philosophy of Pseudoscience: Reconsidering the Demarcation Problem*, edited by Massimo Pigliucci and Maarten Boudry, 345–60. Chicago: University of Chicago Press.

Lambert, W. G. 1960. *Babylonian Wisdom Literature*. Oxford: Clarendon Press.

1975. "The Cosmology of Sumer and Babylon." In *Ancient Cosmologies*, edited by Carmen Blacker and Michael Loewe, 42–65. London: Allen and Unwin.

1998. "The Qualifications of Babylonian Diviners." In *Tikip Santakki Mala Bašmu: Festschrift für Rykle Borger*, edited by Stefan M. Maul, 141–58. Groningen: Styx Publications.

1999–2000. "The Apsû." In *Landscapes: Territories, Frontiers and Horizons in the Ancient Near East. Papers Presented to the XLIV Rencontre Assyriologique Internationale, Venezia, 7–11 July 1997*, edited by L. Milano, S. de Martino, F. M. Fales, and G. B. Lanfranchi, 76–77. Padua: Sargon.

2013. *Babylonian Creation Myths*. Winona Lake, IN: Eisenbrauns.

Lambert, Wilfrid G., Alan R. Millard, and Miguel Civil. 1969. *Atra-hasis: The Babylonian Story of the Flood*. Oxford: Clarendon Press.

Landsberger, Benno. 1926. "Die Eigenbegrifflichkeit der babylonischen Welt." *Islamica* 2: 355–72. Reprinted as *Leistung und Grenze sumerischer und babylonischer Wissenschaft*. With Wolfram von Soden. Darmstadt: Wissenschaftliche Buchgesellschaft, 1974.

1937. *Materialien zum sumerischen Lexikon: Die Serie ana ittišu. Vol. 1*. Rome: Pontifical Biblical Institute.

1957. *Materialien zum sumerischen Lexikon. Vol. 5: The Series HAR-ra = hubullu. Tablets* I–VI. Rome: Pontifical Biblical Institute.

Landsberger, Benno, Erica Reiner, and Miguel Civil. 1970. *Materialien zum sumerischen Lexikon. Vol. 10: The Series HAR-ra = hubullu. Tablets* XVI, XVII, XIX *and Related Texts*. Rome: Pontifical Biblical Institute.

Laporte, Joseph. 2004. *Natural Kinds and Conceptual Change*. Cambridge: Cambridge University Press.

Latour, Bruno. 1990. "Postmodern? No, Simply Amodern! Steps towards an Anthropology of Science." *Studies in History and Philosophy of Science* 21: 145–71.

[1991] 1993. *We Have Never Been Modern*. Translated by Catherine Porter. Cambridge, MA: Harvard University Press.

2004. *Politics of Nature*. Translated by Catherine Porter. Cambridge, MA: Harvard University Press.

2009. "Perspectivism: Type or Bomb?" *Anthropology Today* 25, no. 2: 1–2.

Latour, Bruno, and Steve Woolgar. (1979) 1986. *Laboratory Life: The Construction of Scientific Facts*. 2nd ed. Thousand Oaks, CA: Sage.

Laudan, Larry. (1981) 1984. "A Confutation of Convergent Realism." *Philosophy of Science* 48: 19–49. Republished in Jarrett Leplin, ed., *Scientific Realism*. Berkeley: University of California Press, 1984, 218–49.

1983. "The Demise of the Demarcation Problem." In *The Demarcation between Science and Pseudo-science*, edited by Rachel Lauden, 7–35. Blacksburg: Virginia Tech Center for the Study of Science in Society. Also published in R. S. Cohen and L. Laudan, eds., *Physics, Philosophy and Psychoanalysis: Essays in Honor of Adolf Grünbaum*. Dordrecht: Reidel, 1983, 111–27.

1996. *Beyond Positivism and Relativism*. Boulder, CO: Westview Press.

Lazaris, Stavros, ed. 2020. *A Companion to Byzantine Science*. Leiden: Brill.

Leeten, Lars. 2012. "What Is 'Critique of Worldmaking?' Nelson Goodman's Conception of Philosophy." *Enrahonar* 49: 29–40.

Leggatt, Stuart. 1995. *Aristotle: On the Heavens* I–II. Warminster, PA: Aris & Phillips.

Lehoux, Daryn. 2012. *What Did the Romans Know? An Inquiry into Science and Worldmaking*. Chicago: University of Chicago Press.

2019. "'All Things Are Full of Gods': Naturalism in the Classical World." In *Science without God? Rethinking the History of Scientific Naturalism*, edited by Peter Harrison and Jon H. Roberts, 19–36. Oxford: Oxford University Press.

Leichty, E. 2011. *The Royal Inscription of Esarhaddon, King of Assyria (680–669* BC*)*. Winona Lake, IN: Eisenbrauns.

Lenzi, Alan. 2008. *Secrecy and the Gods: Secret Knowledge in Ancient Mesopotamia and Biblical Israel*. Helsinki: Neo-Assyrian Text Corpus Project.

Lescher, J. H. 1999. "Early Interest in Knowledge." In *The Cambridge Companion to Early Greek Philosophy*, edited by A. A. Long, 225–49. Cambridge: Cambridge University Press.

Levinson, Stephen C. 2003. *Space in Language and Cognition*. Cambridge: Cambridge University Press.

Lévi-Strauss, Claude. (1983) 1992. *The View from Afar*. Chicago: University of Chicago Press.

Lewis, David. (1986) 2001. *On the Plurality of Worlds*. Malden, MA: Wiley-Blackwell.

Lewis, Martin W., and Kären E. Wigen. 1997. *The Myth of Continents: A Critique of Metageography*. Berkeley: University of California Press.

Lindberg, David C. and Robert S. Westman, eds. 1990. *Reappraisals of the Scientific Revolution*. Cambridge: Cambridge University Press.

Lipton, Peter. 2005. "The Medawar Lecture 2004: The Truth about Science." *Philosophical Transactions of the Royal Society B* 360: 1259–69.

Livingstone, Alisdair. 1986. *Mystical and Mythological Explanatory Works of Assyrian and Babylonian Scholars*. Oxford: Clarendon Press.

1989. *Assyrian Court Poetry and Literary Miscellanea*. Helsinki: Helsinki University Press.

2007. "Ashurbanipal: Literate or Not?" *Zeitschrift für Assyriologie* 97: 98–118.

Lloyd, Geoffrey E. R. 1966. *Polarity and Analogy: Two Types of Argumentation in Early Greek Thought*. Cambridge: Cambridge University Press.

1968. *Aristotle: The Growth and Structure of His Thought*. Cambridge: Cambridge University Press.

1970. *Early Greek Science: Thales to Aristotle*. New York: Norton.

1973. *Greek Science after Aristotle*. New York: Norton.

1978. *Aristotle on Mind and the Senses*. Cambridge: Cambridge University Press.

1979. *Magic Reason and Experience: Studies in the Origin and Development of Greek Science*. Cambridge: Cambridge University Press.

1983. *Science, Folklore and Ideology*. Cambridge: Cambridge University Press.

1987. *The Revolutions of Wisdom: Studies in the Claims and Practice of Ancient Greek Science*. Berkeley: University of California Press.

1990. *Demystifying Mentalities*. Cambridge: Cambridge University Press.

1991. *Methods and Problems in Greek Science*. Cambridge: Cambridge University Press.

1992. "Methods and Problems in the History of Ancient Science: The Greek Case." *Isis* 83: 564–77.

1996. *Adversaries and Authorities: Investigations into Ancient Greek and Chinese Science*. Cambridge: Cambridge University Press.

2004. *Ancient Worlds, Modern Reflections: Philosophical Perspectives on Greek and Chinese Science and Culture*. New York: Oxford University Press.

2005. *The Delusions of Invulnerability: Wisdom and Morality in Ancient Greece, China and Today*. London: Duckworth.
2006. *Principles and Practices in Ancient Greek and Chinese Science*. Aldershot: Ashgate.
2007. *Cognitive Variations: Reflections on the Unity and Diversity of the Human Mind*. Oxford: Clarendon Press.
2008. "The Varying Agenda of the Study of the Heavens: Mesopotamia, Greece, China." In *Star Gazing, Fire Phasing, and Healing in China: Essays in Honor of Nathan Sivin*, edited by Michael Nylan, 69–88. Taipei: Academica Sinica, Institute of History and Philology.
2010. "History and Human Nature: Cross-Cultural Universals and Cultural Relativities." *Interdisciplinary Science Reviews* 35: 201–14.
2011. "Multidimensional Reality." *Common Knowledge* 17: 27–30.
2012. *Being, Humanity, and Understanding*. Oxford: Oxford University Press.
2019. "The Clash of Ontologies and the Problems of Translation and Mutual Intelligibility." *HAU: Journal of Ethnographic Theory* 9: 36–43.
2020. *Intelligence and Intelligibility: Cross-Cultural Studies of Human Cognitive Experience*. Oxford: Oxford University Press.
2021a. *Expanding Horizons in the History of Science: The Comparative Approach*. Cambridge: Cambridge University Press.
2021b. "Introduction." *Interdisciplinary Science Reviews* 46: 215–24.

Lloyd, Geoffrey E. R., and Nathan Sivin. 2003. *The Way and the Word: Science and Medicine in Early China and Greece*. New Haven, CT: Yale University Press.

Lloyd, Geoffrey E. R., and Aparecida Vilaça. 2020. *Science in the Forest, Science in the Past*. Chicago: HAU Books. First published *HAU: Journal of Ethnographic Theory* 9 (2019): 36–182.
2023. *Of Jaguars and Butterflies: Metalogues on Issues in Anthropology and Philosophy*. New York: Berghahn Books.

Loison, Laurent. 2016. "Forms of Presentism in the History of Science: Rethinking the Project of Historical Epistemology." *Studies in History and Philosophy of Science Part A* 60: 29–37.

Lynch, Michael. 2013. "Ontography: Investigating the Production of Things, Deflating Ontology." *Social Studies of Science* 43: 444–62.

Magdalino, Paul, and Maria Mavroudi. 2006. *The Occult Sciences in Byzantium*. Geneva: La Pomme d'or.

Magnus, P. D. 2012. *Scientific Enquiry and Natural Kinds: From Planets to Mallards*. New York: Palgrave Macmillan.

Mahner, Martin. 2007. "Demarcating Science from Non-Science." In *Handbook of Philosophy of Science. Vol. 1: General Philosophy of Science—Focal Issues*, edited by Theo A. F. Kuipers, 515–75. Amsterdam: North Holland.
2013. "Science and Pseudoscience: How to Demarcate after the (Alleged) Demise of the Demarcation Problem." In *Philosophy of Pseudoscience: Reconsidering the Demarcation Problem*, edited by Massimo Pigliucci and Maarten Boudry, 29–44. Chicago: University of Chicago Press.

Mak, Bill M. 2013a. "The Date and Nature of Sphujidhvaja's Yavanajātaka Reconsidered in the Light of Some Newly Discovered Materials." *History of Science in South Asia* 1: 1–20.

2013b. "The Last Chapter of Sphujidhvaja's Yavanajātaka Critically Edited with Notes." *SCIAMVS* 14: 59–148.

2014. "The 'Oldest Indo-Greek Text in Sanskrit' Revisited: Additional Readings from the Newly Discovered Manuscript of the Yavanajātaka." *Journal of Indian and Buddhist Studies* 62: 37–41.

2019. "Greco-Babylonian Astral Science in Asia: Patterns of Dissemination and Transformation." In *East–West Encounter in the Science of Heaven and Earth*, edited by Tokimasa Takeda and Bill M. Mak, 14–34. Kyoto: Institute for Research in Humanities, Kyoto University.

Malachowski, Alan, ed. 2020. *A Companion to Rorty*. Hoboken, NJ: Wiley.

Marya, Deepika. 2017. "Postcolonial Literature in the Time of World Literature." In *Postcolonial Gateways and Walls: Under Construction*, edited by Daria Tunca and Janet Wilson, 165–80. Leiden: Brill.

Mates, Benson. 1996. *The Skeptic Way: Sextus Empiricus' Outlines of Pyrrhonism*. Oxford: Oxford University Press.

Maul, Stefan M. 1994. *Zukunftsbewältigung: Eine Untersuchung altorientalischen Denkens anhand der babylonisch-assyrischen Löserituale (Namburbi)*. Mainz: Philipp von Zabern.

2013. *Die Wahrsagekunst im Alten Orient: Zeichen des Himmels und der Erde*. Munich: Verlag C.H. Beck.

McCarty, Willard, Geoffrey E. R. Lloyd, and Aparecida Vilaça, eds. (2021) 2022. *Science in the Forest, Science in the Past: Further Interdisciplinary Explorations*. Milton Park: Routledge.

McCormick, Peter J., ed. 1996. *Starmaking: Realism, Anti-Realism, and Irrealism*. Cambridge, MA: MIT Press.

McKie, Douglas. 1958. *Science and History: An Inaugural Lecture Delivered at University College, London, 22 May 1958*. London: H. K. Lewis.

McMullin, Ernan. 1984a. "The Goals of Natural Science." *Proceedings and Addresses of the American Philosophical Society* 58: 37–54.

1984b. "A Case for Scientific Realism." In *Scientific Realism*, edited by Jarrett Leplin, 8–40. Berkeley: University of California Press.

Minsky, Marvin. 1965. "Matter, Mind, and Models." In *Information Processing 1965: Proceedings of the IFIP [International Federation for Information Processing] Congress, 1965*, edited by W. A. Kalenich. London: MacMillan and Co.

Misiewicz, Zoë. 2018. "The Importance of Experts: Agents in the Transfer of Knowledge between Hellenistic Mesopotamia and the Greek-Speaking World." In "*The Scaffolding of Our Thoughts*": *Essays on Assyriology and the History of Science in Honor of Francesca Rochberg*, edited by C. Jay Crisostomo, Eduardo A. Escobar, Terri Tanaka, and Niek Veldhuis, 317–32. Leiden: Brill.

2019. "John Lydus and Mesopotamian Celestial Omens in Late Antiquity," PhD diss. Institute for the Study of the Ancient World, NYU.

Mitchell, Sandra D. 2002. "Integrative Pluralism." *Biology and Philosophy* 17: 55–70.

Mitchell, W. J. T. 1986. "How Good Is Nelson Goodman?" *Poetics Today* 7: 111–15.

Mizrahi, Moti. 2016. "The History of Science as a Graveyard of Theories: A Philosophers' Myth?" *International Studies in Philosophy of Science* 30, no. 3: 263–78.

Mizrahi, Moti, ed. 2018. *The Kuhnian Image of Science: Time for a Decisive Transformation?* London: Rowman & Littlefield.

Momigliano, Arnaldo. 1990. *Alien Wisdom: The Limits of Hellenization.* Cambridge: Cambridge University Press.

Montelle, Clemency. 2011. *Chasing Shadows: Mathematics, Astronomy, and the Early History of Eclipse Reckoning.* Baltimore, MD: Johns Hopkins University Press.

Montelle, Clemency, and Kim Plofker. 2018. *Sanskrit Astronomical Tables.* Cham: Springer Nature.

Morenz, L. D. 2003. "Neuassyrische visuell-poetische Bilder-Schrift und ihr Vor-Bild." In *Herrscherpräsentation und Kulturkontake. Ägypten–Levante–Mesopotamien*, edited by L. D. Morenz and E. Bosshard-Nepustil, 197–229. Münster: Ugarit Verlag.

Morrison, Margaret. 2007. "Where Have All the Theories Gone?" *Philosophy of Science* 74: 195–228.

Neugebauer, Otto. 1942. "On Some Astronomical Papyri and Related Problems of Ancient Geography." *Transactions of the American Philosophical Society* 32, no. 2: 251–63.

1945. "The History of Ancient Astronomy: Problems and Methods." *Journal of Near Eastern Studies* 4: 1–38.

1947. "A Table of Solstices from Uruk." *Journal of Cuneiform Studies* 1: 143–48.

1951. "The Study of Wretched Subjects." *Isis* 42: 111.

(1955) 1983. *Astronomical Cuneiform Texts.* 3 vols. 2nd ed. London: Lund Humphries.

1963. "The Survival of Babylonian Methods in the Exact Sciences of Antiquity and Middle Ages." *Proceedings of the American Philosophical Society* 107: 528–35.

1969. *The Exact Sciences in Antiquity.* New York: Dover.

1975. *A History of Ancient Mathematical Astronomy.* 3 vols. Berlin: Springer.

1988. "A Babylonian Lunar Ephemeris from Roman Egypt." In *A Scientific Humanist: Studies in Memory of Abraham Sachs*, edited by Erle Leichty, Maria DeJ. Ellis, and Pamela Gerardi, 301–4. Philadelphia: University Museum.

1989. "From Assyriology to Renaissance Art." *Proceedings of the American Philosophical Society* 133: 391–403.

Neugebauer, O., and H. B. van Hoesen. 1959. *Greek Horoscopes.* Philadelphia: American Philosophical Society.

Nickles, Thomas. 2013. "The Problem of Demarcation." In *Philosophy of Pseudoscience: Reconsidering the Demarcation Problem*, edited by

Massimo Pigliucci and Maarten Boudry, 101–20. Chicago: University of Chicago Press.
Noegel, Scott B. 2017. "God of Heaven and Sheol: The 'Unearthing' Creation." *Hebrew Studies* 58: 119–44.
North, John D. 1998. "Olaf Pedersen (1920–1997)." *Journal for the History of Astronomy* 29, no. 2: 212–14.
Oldfather, C. H. 1998. *Diodorus Siculus, Library of History Books* 1–11.34. Cambridge, MA: Harvard University Press.
Olen, Peter. 2019. "Ajdukiewicz and Sellars on World-Perspectives." In *Wilfrid Sellars and Twentieth-Century Philosophy*, edited by Stefan Brandt and Anke Breunig, 55–74. New York: Routledge.
Oliveira, J. C. Pinto de. 2012. "Kuhn and the Genesis of the 'New Historiography of Science.'" *Studies in History and Philosophy of Science* 43: 115–21.
Oppenheim, A. Leo. 1960. "Assyriology – Why and How?" *Current Anthropology* 1: 409–23.
(1965) 1977. *Ancient Mesopotamia: Portrait of a Dead Civilization*. Rev. ed. Chicago: University of Chicago Press.
1967. *Letters from Mesopotamia*. Chicago: University of Chicago Press.
1974. "A Babylonian Diviner's Manual." *Journal of Near Eastern Studies* 33: 197–220.
Ossendrijver, Mathieu. 2011a. "Exzellente Netzwerke: Die Astronomen Von Uruk." In *The Empirical Dimension of Ancient Near Eastern Studies*, edited by Gebhard J. Selz and Klaus Wagensonner, 631–44. Vienna: LIT Verlag.
2011b. "Science in Action: Networks in Babylonian Astronomy." In *Proceedings of the Conference "Babylon–Wissenskultur Zwischen Orient und Okzident," Pergamon Museum, Berlin 26–28/6/2008*, edited by Eva Cancik-Kirschbaum, Margarete van Ess, and Joachim Marzahn, 213–21. Berlin: De Gruyter.
2012. *Babylonian Mathematical Astronomy: Procedure Texts*. New York: Springer.
2021a. "Astral Science in Uruk during the First Millennium BCE: Libraries, Communities and Transfer of Knowledge." In *Uruk-altorientalische Metropole und Kulturzentrum: 8. Internationales Colloquium der Deutschen Orient-Gesellschaft 25. und 26. April 2013, Berlin*, edited by Margarete van Ess, 319–42. Wiesbaden: Harrassowitz Verlag.
Ossendrijver, M., ed. 2021b. "Scholars, Priests, and Temples: Babylonian and Egyptian Science in Context." Special Issue, *JANEH* 8, nos. 1–2.
Ossendrijver, Mathieu. 2024. *Babylonian Mathematical Astronomy: Planetary Tables*. Berlin: Springer Verlag.
Paleček, Martin, and Mark Risjord. 2012. "Relativism and the Ontological Turn within Anthropology." *Philosophy of the Social Sciences* 43: 3–23.
Panaino, Antonio C. D. 2021. *A Walk through the Iranian Heavens: Spherical and Non-Spherical Cosmographic Models in the Imagination of Ancient Iran and Its Neighbors*. Leiden: Brill.
Park, Peter K. J. 2013. *Africa, Asia, and the History of Philosophy: Racism in the Formation of the Philosophical Canon 1780–1830*. Albany, NY: SUNY Press.

Parpola, Simo. 1970. *Letters from Assyrian Scholars to the Kings Esarhaddon and Assurbanipal*, Vol. 1. Kevelaer: Butzon & Bercker.

1993. *Letters from Assyrian and Babylonian Scholars*. Helsinki: Helsinki University Press.

Pedersen, Moten Axel. 2012. "Common Sense: A Review of Certain Reviews of the 'Ontological Turn.'" *Anthropology of This Century* 5. http://aotcpress.com/articles/common_nonsense/#sthash.1yvGKe2k. dpuf.

Pedersen, Olaf. (1974) 1993. *Early Physics and Astronomy*. 2nd ed. Cambridge: Cambridge University Press.

Pickering, Andrew. 2017. "The Ontological Turn: Taking Different Worlds Seriously." Special Issue: Multiple Nature-Cultures, Diverse Anthropologies. *Social Analysis: The International Journal of Anthropology* 61: 134–50.

Pigliucci, Massimo. 2013. "The Demarcation Problem: A (Belated) Response to Laudan." In *Philosophy of Pseudoscience: Reconsidering the Demarcation Problem*, edited by Massimo Pigliucci and Maarten Boudry, 9–28. Chicago: University of Chicago Press.

Pigliucci, Massimo, and Maarten Boudry, eds. 2013. *Philosophy of Pseudoscience: Reconsidering the Demarcation Problem*. Chicago: University of Chicago Press.

Pina-Cabral, João de. 2014a. "World: An Anthropological Examination (part 1)." *Journal of Ethnographic Theory* 4: 49–73.

2014b. "World: An Anthropological Examination (part 2)." *HAU: Journal of Ethnographic Theory* 4: 149–84.

Pingree, David, ed. 1974. *Hephaestionis Thebani Apotelesmaticorum Libri Tres*. Leipzig: B.G. Teubner.

1976. *Dorothei Sidonii Carmen Astrologicum*. Leipzig: B.G. Teubner.

Pingree, David. 1978. *The Yavanajātaka of Sphujidhvaja*. Cambridge, MA: Harvard University Press.

1982. "Mesopotamian Astronomy and Astral Omens in Other Civilizations." In *Mesopotamien und seine Nachbarn: Politische und kulturelle Wechselbezeihungen im Alten Vorderasien vom 4. bis 1. Jahrtausend v. Chr*, edited by Hans Nissen and Johannes Renger, 613–31. Berlin: Reimer.

Pingree, David, ed. 1986. *Vettii Valentis Antiocheni Anthologiarum Libri Novem*. Leipzig: B.G. Teubner.

Pingree, David. 1987. "Venus Omens in India and Babylon." In *Language, Literature, and History: Philological and Historical Studies Presented to Erica Reiner*, edited by Francesca Rochberg-Halton, 293–315. New Haven, CT: American Oriental Society.

1992a. "Hellenophilia versus the History of Science." *Isis* 83: 554–63.

1992b. "Mesopotamian Omens in Sanskrit." In *La circulation des biens, des personnes et des idées dans le Proche-Orient ancien: Actes de la XXXVIIIe Rencontre Assyriologique Internationale, Paris, 8–10 juillet 1991*, edited by Dominique Charpin and Francis Joannès, 375–79. Paris: Edité par Editions Recherche sur les Civilisations.

1997. *From Astral Omens to Astrology: From Babylon to Bikaner*. Rome: Istituto Italiano per l'Africa e l'Oriente.

1998. "Legacies in Astronomy and Celestial Omens." In *The Legacy of Mesopotamia*, edited by Stephanie Dalley, 125–37. Oxford: Oxford University Press.

2003. "The Logic of Non-western Science: Mathematical Discoveries in Medieval India." *Daedalus* 132: 45–53.

Pisano, Raffaela, Joseph Agassi, and Daria Drozdova, eds. 2018. *Hypotheses and Perspectives in the History and Philosophy of Science: Homage to Alexandre Koyré 1892–1964*. Cham: Springer Nature.

Plofker, Kim. 2008. *Mathematics in India: 500 BCE–1800 CE*. Princeton, NJ: Princeton University Press.

Pope, Alexander. 1905. *An Essay on Man: Moral Essays and Satires*. With an Introduction by Henry Morley. London: Cassell.

Popper, Karl. (1935) 2002. *The Logic of Scientific Discovery*. Vienna: Julius Springer. [Translated by Hutchinson. English translation: 1959; Routledge 1992. Reissued by Routledge 2002].

(1953) 2002. *The Logic of Scientific Discovery*. London: Routledge.

(1963) 2004. *Conjectures and Refutations*. London: Routledge.

Putnam, Hilary. 1981. *Reason, Truth and History*. Cambridge: Cambridge University Press.

1982. "Why There Isn't a Ready-Made World." *Synthese* 51: 141–67.

(1987) 1989. "Truth and Convention: On Davidson's Refutation of Conceptual Relativism." *Dialectica* 41: 69–77.Republished Michael Krausz, ed., *Relativism: Interpretation and Confrontation*. Notre Dame, IN: University of Notre Dame Press, 1989, 173–82.

Quack, Joachim. 2018a. *Eine magische Stele aus dem Badischen Landesmuseum Karlsruhe (Inv. H 1049)*. Heidelberg: Heidelberger Akademie der Wissenschaften.

2018b. "Egypt as Astronomical Astrological Centre between Mesopotamia and India." In *The Interactions of Ancient Astral Science*, by David Brown, with contributions by Jonathan Ben-Dov, Harry Falk, Geoffrey Lloyd, Raymond Mercier, Antonio Panaino, Joachim Quack, Alexandra von Lieven, and Michio Yano, 69–123. Bremen: Hempen.

2022. "Ein frühdemotischer Weisheitstext: Papyrus Berlin 23504." In *One Who Loves Knowledge: Studies in Honor of Richard Jasnow*, edited by Betsy Bryan, Mark Smith, Christina di Cerbo, Marina Escolano-Poveda, and Jill S. Waller, 323–33. Columbus, GA: Lockwood Press.

Quine, Willard von Orman. 1948. "On What There Is." *The Review of Metaphysics* 2: 21–38.

1951. "Main Trends in Recent Philosophy: Two Dogmas of Empiricism." *The Philosophical Review* 60: 20–43.

1953. *From a Logical Point of View*. New York: Harper.

1968. "Ontological Relativity." *The Journal of Philosophy* 65: 185–212.

1969. *Ontological Relativity and Other Essays*. New York: Columbia University Press.

1992. *Pursuit of Truth*. Cambridge, MA: Harvard University Press.

Radner, Karen, and Eleanor Robson. 2011. *The Oxford Handbook of Cuneiform Culture*. Oxford: Oxford University Press.

Rawlinson, Henry C., ed. 1875. *The Cuneiform Inscriptions of Western Asia. Vol. IV: A Selection from the Miscellaneous Inscriptions of Assyria*. London: The British Museum.

Reiner, Erica. 1995. *Astral Magic in Babylonia*. Philadelphia: American Philosophical Society.

Reiner, Erica, and David Pingree. 1975. *The Venus Tablet of Ammiœaduqa: Babylonian Planetary Omens: Part* 1. Malibu, CA: Undena Publications.

1981. *Enūma Anu Enlil, Tablets 50–51*. Malibu, CA: Undena Publications.

1998. *Babylonian Planetary Omens: Part 3*. Groningen: Styx Publications.

2005. *Babylonian Planetary Omens: Part 4*. Leiden: Brill.

Ricoeur, Paul. 1980. "Critical Discussions: Ways of Worldmaking." *Philosophy and Literature* 4: 107–20.

Roaf, Michael, and Annette Zgoll. 2001. "Assyrian *Astroglyphs*: Lord Aberdeen's Black Stone and the Prisms of *Esarhaddon*." *Zeitschrift für Assyriologie* 91: 264–95.

Robbins, F. E. 1971. *Ptolemy Tetrabiblos*. Cambridge, MA and London: Harvard University Press and William Heinemann.

Robson, Eleanor. 1999. *Mesopotamian Mathematics 2100–1600 BC: Technical Constants in Bureaucracy and Education*. Oxford: Clarendon Press.

2001. "The Uses of Mathematics in Ancient Iraq (6000–600 BC)." In *Mathematics across Cultures: The History of Non-Western Mathematics*, edited by Helaine Selin, 93–114. Dordrecht: Kluwer Academic.

2011. "The Production and Dissemination of Scholarly Knowledge." In *The Oxford Handbook of Cuneiform Culture*, edited by Karen Radner and Eleanor Robson, 557–76. Oxford: Oxford University Press.

2013. "Reading the Libraries of Assyria and Babylonia." In *Ancient Libraries*, edited by Jason König, Katerina Oikonomopoulou, and Greg Woolf, 37–56. Cambridge: Cambridge University Press.

2019. *Ancient Knowledge Networks: A Social Geography of Cuneiform Scholarship in First-Millennium Assyria and Babylonia*. London: UCL Press.

Robson, Eleanor, and Kathryn Stevens. 2019. *Scholarly Tablet Collections in First-Millennium Assyria and Babylonia, c.700–200 BCE*. In *Libraries before Alexandria: Ancient Near Eastern Traditions*, edited by Kim Ryholt and Gojko Barjamovic, 319–366. Oxford: Oxford University Press.

Rochberg, Francesca. 1992. "The Cultures of Ancient Science: Some Historical Reflections." *Isis* 83: 547–53.

1993. "The *Cultural Locus of Astronomy* in Late Babylonia." In *Die Rolle der Astronomie in den Kulturen Mesopotamiens. Beiträge zum 3. Grazer Morgenländischen Symposion (23–27. September 1991)*, edited by Hans D. Galter, 31–45. Graz: GrazKult.

1996. "Personifications and Metaphors in Babylonian Celestial Omina." *Journal of the American Oriental Society* 116: 475–85.

1998. *Babylonian Horoscopes*. Philadelphia: American Philosophical Society.

2000. "Scribes and Scholars: The Tupšār Enūma Anu Enlil." In *Assyriologica et Semitica: Festschrift für Joachim Oelsner anläßich aeines 65. Geburtstages am 18. Februar 1997*, edited by Joachim Oelsner, Joachim Marzahn, and Hans Neumann, 359–75. Münster: Ugarit Verlag.

2004. *The Heavenly Writing: Divination, Horoscopy and Astronomy in Mesopotamian Culture*. Cambridge: Cambridge University Press.

2009a. "Inference, Conditionals, and Possibility in Ancient Mesopotamian Science." *Science in Context* 22: 4–25.

2009b. "The Stars Their Likenesses." In *What Is a God? Anthropomorphic and Non-Anthropomorphic Aspects of Deity in Ancient Mesopotamia*, edited by Barbara Nevling Porter, 1–52. Winona Lake, IN: Eisenbrauns.

2010a. *In the Path of the Moon*. Leiden: Brill.

2010b. "If P, Then Q: Form, Reasoning and Truth in Babylonian Divination." In *Divination and Interpretation of Signs in Antiquity*, edited by Amar Annus, 19–27. Chicago: Oriental Institute Publications.

2010c. "Sheep and Cattle, Cows and Calves: The Sumero-Akkadian Astral Gods as Livestock." In *Opening the Tablet Box: Near Eastern Studies in Honor of Benjamin R. Foster*, edited by Sarah Melville and Alice Slotsky, 347–59. Leiden: Brill.

2011. "Divine Causality and Cuneiform Divination." In *A Common Cultural Heritage: Studies in Mesopotamia and the Biblical World in Honor of Barry L. Eichler*, edited by Grant Frame, Erle Leichty, Jeffery Tigay, and Steve Tinney, 279–94. Bethesda, MD: CDL Press.

2015. "The Babylonians and the Rational: Contexts of Rational Reasoning in Cuneiform Scribal Scholarship." In *In the Wake of the Compendia: How Technical Handbooks and Encyclopedia Reshape Early Mesopotamian Empiricism, Science, Technology and Medicine in Ancient Cultures*, edited by Cale Johnson, 209–48. Berlin: De Gruyter.

2016a. *Before Nature: Cuneiform Knowledge and the History of Science*. Chicago: University of Chicago Press.

2016b. "A Critique of the Cognitive-Historical Thesis of *The Intellectual Adventure*." In *The Adventure of the Human Intellect: Self, Society and the Divine in Ancient World Cultures*, edited by Kurt A. Raaflaub, 16–28. Malden, MA: Wiley-Blackwell.

2016c. "Canon and Power in Cuneiform Scribal Scholarship." In *Problems of Canonicity and Identity Formation in Ancient Egypt and Mesopotamia*, edited by Kim Ryholt and Goyko Barjamovic, 217–29. Copenhagen: Museum Tusculanum Press.

2018a. "The Near Eastern Heritage in Greco-Roman Astronomy." In *Studying the Near and Middle East at the Institute for Advanced Study, Princeton, 1935–2018*, edited by Sabine Schmidtke, 11–20. Piscataway, NJ: Gorgias Press.

2018b. "The Catalogues of *Enūma Anu Enlil*." In *Assyrian and Babylonian Text Catalogues: Medicine, Magic and Divination*, edited by Ulrike Steinert, 121–36. Boston: De Gruyter.

2018c. "Reasoning, Representing, and Modeling in Babylonian Astronomy." *Journal of Ancient Near Eastern History* 5: 131–47.

2019. "Categories, Kinds and Determinatives." In *Encoding Metalinguistic Awareness: Ancient Mesopotamia and Beyond*, edited by Eva Cancik-Kirschbaum and J. Cale Johnson, 53–64. Berlin: Pewe-Verlag.

2020. "Mesopotamian Cosmology." In *Blackwell's Companion to the Ancient Near East*, edited by Daniel Snell, 305–20. Oxford: Blackwell.

2021. "Mereological Themes in Cuneiform Worldmaking." *Interdisciplinary Sciences Reviews* 46: 386–404. Reprinted in Willard McCarty, Geoffrey E. R. Lloyd, and Aparecida Vilaça, eds., *Science in the Forest, Science in the Past: Further Interdisciplinary Explorations*, 169–87. Milton Park: Routledge, 2022.

2022. "Cuneiform Classification: Worldmaking, Science, and the Ontological Turn." *Mètis* 20: 17–33.

Rochberg-Halton, Francesca. 1983. "Stellar Distances in Early Babylonian Astronomy: A New Perspective on the Hilprecht Text (HS 229)." *Journal of Near Eastern Studies* 42: 209–17.

1984. "New Evidence for the History of Astrology." *Journal of Near Eastern Studies* 43: 115–40.

1988a. *Aspects of Babylonian Celestial Divination: The Lunar Eclipse Tablets of Enūma Anu Enlil*. Horn: Berger & Söhne.

1988b. "Elements of the Babylonian Contribution to Hellenistic Astrology." *Journal of the American Oriental Society* 108: 51–62.

1988c. "Benefic and Malefic Planets in Babylonian Astrology." In *A Scientific Humanist: Studies in Memory of Abraham Sachs*, edited by Erle Leichty, Maria de Jong Ellis, and Pamela Gerardi, 323–28. Philadelphia: University Museum, University of Pennsylvania.

1989. "Babylonian Seasonal Hours." *Centaurus* 32: 146–70.

1991. "Between Observation and Theory in Babylonian Astronomical Texts." *Journal of Near Eastern Studies* 50: 107–20.

Rorty, Richard. 1972. "The World Well Lost." *The Journal of Philosophy* 69: 649–65.

1979. *Philosophy and the Mirror of Nature*. Princeton, NJ: Princeton University Press.

1986. "On Ethnocentrism: A Reply to Clifford Geertz." *Michigan Quarterly Review* 25: 525–34.

1998. *Truth and Progress, Philosophical Papers*. New York: Cambridge University Press.

Rosenfeld, L. 1953. "Introduction." *Centaurus: International Magazine of the History of Mathematics, Science, and Technology* 3: 7–11.

Rouse, J. 1991. "Philosophy of Science and the Persistent Narratives of Modernity." *Studies in History and Philosophy of Science* 22: 141–62.

1992. "What Are Cultural Studies of Scientific Knowledge?" *Configurations* 1: 57–94.

Rumor, Maddalena. 2021. "Babylonian Astro-Medicine, Quadruplicities and Pliny the Elder." *Zeitschrift für Assyriologie und Vorderasiatische Archäologie* 111: 47–76.

Rutz, M. 2013. *Bodies of Knowledge in Ancient Mesopotamia: The Diviners of Late Bronze Age Emar and Their Tablet Collection*. Leiden: Brill.

Sachs, Abraham J. 1955. *Late Babylonian Astronomical and Related Texts Copied by T. G. Pinches and J. N. Strassmaier*. With the cooperation of J. Schaumberger. Providence, RI: Brown University Press.

Sachs, Abraham J., and Hermann Hunger. 1988. *Astronomical Diaries and Related Texts from Babylonia. Vol. 1: Diaries from 652 B. C. to 262 B. C.* Vienna: Verlag der Österreichischen Akademie der Wissenschaften.

1989. *Astronomical Diaries and Related Texts from Babylonia. Vol. 2: Diaries from 261 B. C. to 165 B. C.* Vienna: Verlag der Österreichischen Akademie der Wissenschaften.

1996a. *Astronomical Diaries and Related Texts from Babylonia. Vol. 3: Diaries from 164 B. C. to 61 B. C.* Vienna: Verlag der Österreichischen Akademie der Wissenschaften.

1996b. *Astronomical Diaries and Related Texts from Babylonia. Vol. 6: Goal Year Texts.* Vienna: Verlag der Österreichischen Akademie der Wissenschaften.

Sankey, Howard. 2000. "Kuhn's Ontological Relativism." *Science and Education* 9: 59–75.

2018. "The Demise of the Incommensurability Thesis." In *The Kuhnian Image of Science: Time for a Decisive Transformation?*, edited by Moti Mizrahi, 75–91. London: Rowman & Littlefield.

Sarton, George. 1924. "Review of Lynn Thorndike. *A History of Magic and Experimental Science during the First Thirteen Centuries of Our Era, 1923.*" *Isis* 6: 74–89.

1950. "Review of Ethel Stefana Drower, The Book of the Zodiac (Sfar Malwasia)." *Isis* 41: 374.

Schaumberger, J. 1952. "Die Ziqpu-Gestirne nach neuen Keilschrifttexten." *Zeitschrift fur Assyriologie* 50: 214–29.

1955. "Anaphora und Aufganskalender in neuen Ziqpu-Texten." *Zeitschrift fur Assyriologie* 51: 237–51.

Scheffler, Israel. 1967. *Science and Subjectivity.* Indianapolis, IN: Bobbs-Merrill.

1980. "The Wonderful Worlds of Nelson Goodman." *Synthese* 45: 201–9. Republished Peter McCormick, ed., *Starmaking: Realism, Anti-realism, and Irrealism.* Cambridge, MA: MIT Press, 1996, 133–42.

1996. "Worldmaking: Why Worry." In *Starmaking: Realism, Anti-realism, and Irrealism*, edited by Peter J. McCormick, 171–77. Cambridge, MA: MIT Press.

2000. "A Plea for Plurealism." *Erkenntnis* 52: 161–73.

2001. "My Quarrels with Nelson Goodman." *Philosophy and Phenomenological Research* 62: 665–77.

2009. *Worlds of Truth: A Philosophy of Knowledge.* Malden, MA: Wiley-Blackwell.

Schmidtchen, Eric. 2018. "Esagil-kīn-apli's Catalogue of Sakikkû and Alamdimmû." In *Assyrian and Babylonian Scholarly Text Catalogues*, edited by Ulrike Steinert, 137–57. Berlin: Walter de Gruyter.

2019. "The Series *Šumma Ea liballiṭka* Revisited." In *Visualizing the Invisible with the Human Body: Physiognomy and Ekphrasis in the Ancient World*, edited by J. Cale Johnson and Alessandro Stavru, 81–118. Berlin: Walter de Gruyter.

Schrödinger, Erwin. 1996. *Nature and the Greeks and Science and Humanism.* With a foreword by Roger Penrose. 2nd ed. Cambridge: Cambridge University Press.

Schuster-Brandis, Anais. 2008. *Steine als Schutz- und Heilmittel: Untersuchung zu ihrer Verwendung in der Beschwörungskunst Mesopotamiens im 1. Jt. v. Chr.* Münster: Ugarit Verlag.

Schwemer, Daniel. 2019. "Mesopotamia." In *Guide to the Study of Ancient Magic*, edited by David Frankfurter, 36–64. Leiden: Brill.

Scurlock, Jo Ann. 1997. "Assyrian Hieroglyphs Enhanced." *N.A.B.U.* 1997/92, 85–86.

2006. *Magico-Medical Means of Treating Ghost-Induced Illnesses in Ancient Mesopotamia*. Leiden: Brill.

Scurlock, Jo Ann, and Burton Andersen. 2005. *Diagnoses in Assyrian and Babylonian Medicine: Ancient Sources, Translations, and Modern Medical Analyses*. Urbana: University of Illinois Press.

Selz, Gebhard J. 2023. "Animal Categorization in Mesopotamia and the Origins of Natural Philosophy." In *Seen Not Heard: Composition, Iconicity and the Classifier Systems in Logosyllabic Scripts*, edited by Ilona Zsolnay, 91–114. Chicago: Institute for the Study of Ancient Cultures of the University of Chicago Publications.

Severi, Carlo. 2013. "Philosophies without Ontology." *HAU: Journal of Ethnographic Theory* 3: 192–96.

2015. "Transmutating Beings: A Proposal for an Anthropology of Thought." In *Translating Worlds: The Epistemological Space of Translation*, edited by Carlo Severi and William F. Hanks, 51–88. Chicago: HAU Books.

Shweder, Richard A. 2000. "Review of Clifford Geertz, *Available Light*." *Science* 24: 1511–12.

Siegel, Harvey. 1984. "Relativism, Realism, and Rightness: Notes on Goodmanian Worldmaking." *Journal of Thought* 19: 16–35.

Sismondo, Sergio. 2015. "Ontological turns, turnoffs and roundabouts." *Social Studies of Science* 45: 441–48.

Sivin, Nathan. 2005. "A Multi-dimensional Approach to Research on Ancient Science." *East Asian Science Technology and Medicine* 23: 10–25.

Skinner, Quentin. 1969. "Meaning and Understanding in the History of Ideas." *History and Theory* 8: 3–53.

Stadhouders, Henry. 2011. "The Pharmacopoeial Handbook Šammu Šikinšu: An Edition." *Journal des Médecines Cunéiformes* 18: 3–51.

2012. "The Pharmacopoeial Handbook Šammu Šikinšu: A Translation." *Journal des Médecines Cunéiformes* 19: 1–20.

Starr, Ivan. 1983. *The Rituals of the Diviner*. Malibu, CA: Undena.

Steele, John M. 2000a. *Observations and Predictions of Eclipse Times by Early Astronomers*. Dordrecht: Kluwer Academic.

2000b. "Eclipse Prediction in Mesopotamia." *Archive for History of Exact Sciences* 54: 421–54.

2007a. "The Length of the Month in Mesopotamian Calendars of the First Millennium BC." In *Calendars and Years: Astronomy and Time in the Ancient Near East*, edited by John M. Steele, 133–48. Oxford: Oxbow Books.

2007b. "Celestial Measurement in Babylonian Astronomy." *Annals of Science* 64: 293–325.

2011a. "Making Sense of Time: Observational and Theoretical Calendars." In *The Oxford Handbook of Cuneiform Culture*, edited by Karen Radner and Eleanor Robson, 470–85. Oxford: Oxford University Press.

2011b. "Goal-Year Periods and Their Use in Predicting Planetary Phenomena." In *The Empirical Dimension of Ancient Near Eastern Studies*, edited by Gebhard J. Selz and Klaus Wagensonner, 101–10. Berlin: LIT Verlag.

2014. "Late Babylonian Ziqpu-Star Lists: Written or Remembered Traditions of Knowledge?" In *Traditions of Written Knowledge in Ancient Egypt and Mesopotamia. Proceedings of Two Workshops Held at Goethe-University, Frankfurt/Main in December 2011 and May 2012*, edited by Daliah Bawanypeck and Annette Imhausen, 123–51. Münster: Ugarit Verlag.

2015. "A Late Babylonian Compendium of Calendrical and Stellar Astrology." *Journal of Cuneiform Studies* 67: 187–215.

2019. "Astronomical Activity in the 'House of the Asipus' in Uruk." In *Scholars and Scholarship in Late Babylonian Uruk*, edited by Christine Proust and John Steele, 147–70. Cham: Springer Nature.

2022. "An Alternative Interpretation of BM 76829: Astrological Schemes for Length of Life and Parts of the Body." *Archive for History of Exact Sciences* 76: 1–14.

Steele, John M., and Annette Imhausen. 2002. *Under One Sky. Astronomy and Mathematics in the Ancient Near East*. Münster: Ugarit Verlag.

Steinert, Ulrike, ed. 2018. *Assyrian and Babylonian Text Catalogues: Medicine, Magic and Divination*. Berlin: De Gruyter.

Steinkeller, Piotr. 2005. "Of Stars and Men: The Conceptual and Mythological Setup of Babylonian Extispicy." In *Biblical and Oriental Essays in Memory of William L. Moran*, edited by Augustinus Gianto, 11–47. Rome: Pontifical Biblical Institute.

Strathern, Marilyn. 1980. "No Nature, No Culture: The Hagen Case." In *Nature, Culture and Gender*, edited by Carol MacCormack and Marilyn Strathern, 174–222. Cambridge: Cambridge University Press.

2004. *Partial Connections*. Walnut Creek, CA: Altamira Press.

Stevens, Kathryn. 2019. *Between Greece and Babylonia: Hellenistic Intellectual History in Cross-Cultural Perspective*. Cambridge: Cambridge University Press.

Stroud, Barry. 2004. "The Charm of Naturalism." In *Naturalism in Question*, edited by Mario de Caro and David MacArthur, 21–35. Cambridge, MA: Harvard University Press.

Sturman, Siep. 2021. "Ethnocentrism in Retrospect: A Response to Jonathan Benthall." *Anthropology Today* 37: 22–23.

Sumner, William Graham. 1906. *Folkways: A Study of the Sociological Importance of Usages, Manners, Customs, Mores, and Morals*. New York: New American Library.

Swerdlow, Noel M. 1993. "Montucla's Legacy: The History of the Exact Sciences." *Journal of the History of Ideas* 54: 299–328.

1998. *The Babylonian Theory of the Planets*. Princeton, NJ: Princeton University Press.

Tahko, Tuomas E. 2021. *Unity of Science*. Cambridge: Cambridge University Press.

Tallqvist, Knut. 1938. *Akkadische Götterepitheta*. Helsinki: Societas Orientalis Fennica.

Thoren, Victor. 1991. "Review of *Centaurus: International Magazine of the History of Mathematics, Science, and Technology*, edited by Kirsti Andersen, Ole Knudsen, Kurt Møller Pedersen, and Olaf Pedersen." *Isis* 82: 304.

Thorndike, Lynn, 1923–58. *A History of Magic and Experimental Science*. 8 vols. New York: Columbia University Press.

Tinney, Steve. 2011. "Tablets of Schools and Scholars: A Portrait of the Old Babylonian Corpus." In *The Oxford Handbook of Cuneiform Culture*, edited by Karen Radner and Eleanor Robson, 577–96. Oxford: Oxford University Press.

Tola, Florencia, and Antonela dos Santos. 2020. "Ontology and Ontologies: Theoretical, Political, and Methodological Debates." *América Crítica* 4: 163–72.

Tola, Florencia, and Celeste Medrano. 2020. "When Histories and Ontologies Compose Multiple Worlds: The Past among the Toba People of the Argentinean Chaco." *América Crítica* 4: 25–34.

Toomer, Gerald J. 1984. *Ptolemy's Almagest*. Princeton, NJ: Princeton University Press.

Tosh, Nick. 2003. "Anachronism and Retrospective Explanation: In Defense of a Present-Centred History of Science." *Studies in History and Philosophy of Science* 34: 647–59.

Tugendhaft, Aaron. 2019. "Archaeology and Jihad: Baron Max von Oppenheim at Tell Halaf." http://www.cabinetmagazine.org/kiosk/tugendhaft_aaron_29_october_2019.php.

2020. *The Idols of Isis*. Chicago: University of Chicago Press.

van de Mieroop, Marc. 1997. "On Writing a History of the Ancient Near East." *Bibliotheca Orientalis* 54: 285–305.

2016. *Philosophy before the Greeks: The Pursuit of Truth in Ancient Babylonia*. Princeton, NJ: Princeton University Press.

Van der Spek, R. J., Jan Luiten van Zanden, and Bas van Leeuwen eds. 2014. *A History of Market Performance: From Ancient Babylonia to the Modern World*. London: Routledge.

van der Waerden, B. L. (1954) 1988. *Science Awakening I*. Leiden: Noordhoff International Publishing.

1974. *Science Awakening* II*: The Birth of Astronomy*. Leiden: Noordhoff International Publishing.

van Fraassen, Bas C. 1995. "'World' Is Not a Count Noun." *Noûs* 29: 139–57.

van Heur, Bas, Loet Leydesdorff, and Sally Wyatt. 2012. "Turning to Ontology in STS? Turning to STS through 'Ontology'?" *Social Studies of Science* 43: 341–62.

van Soldt, Wilfrid H. 1995. *Solar Omens of Enuma Anu Enlil: Tablets 23(24)–29(30)*. Leiden: Nederlands Historisch-Archaeologisch Instituut.

Veenhof, Klaas R. 2021. "Once More ana e/ittišu, Especially in Old Assyrian (OA)." *Zeitschrift für altorientalische und biblische Rechtsgeschichte* 27: 17–30.

Veldhuis, Niek. 2014. *History of the Cuneiform Lexical Tradition*. Münster: Ugarit Verlag.

Venkatesan, Soumhya, ed. 2010. "*Ontology Is Just Another Word for Culture*: Motion Tabled at the 2008 Meeting of the Group for Debates in Anthropological Theory." *Critique of Anthropology* 30: 152–200.

Verderame, L. 2002. *Le Tavole* I–VI *della Serie Astrologica Enūma Anu Enlil*. Messina: DiScAM.

Virolleaud, Charles. 1908. *L'Astrologie chaldéenne: le livre intitulé "Enuma (Anu ilu) Bel"*. Paris: Librairie Paul Geuthner.

Viveiros de Castro, Eduardo. 1992. *Enemy's Point of View: Humanity and Diversity in an Amazonian Society*. Translated by Catherine V. Howard. Chicago: University of Chicago Press.

1998. "Cosmological Deixis and Amerindian Perspectivism." *Journal of Royal Anthropological Institute* 4, no. 3: 469–88.

2003. "And: After-Dinner Speech Given at *Anthropology and Science*. 5th Decennial Conference of the Association of Social Anthropologists of Great Britain and Commonwealth, 14 July 2003." *Manchester Papers in Social Anthropology* 7: 1–20.

2004a. "Perspectival Anthropology and the Method of Controlled Equivocation." *Tipití: Journal of the Society for the Anthropology of Lowland South America* 2, no. 1: 3–22.

2004b. "Exchanging Perspectives: The Transformation of Objects into Subjects in Amerindian Ontologies." *Common Knowledge* 10: 463–84. Republished *Common Knowledge* 25 (2019): 21–42.

2012. *Cosmological Perspectivism in Amazonia and Elsewhere*. Manchester: HAU Network of Ethnographic Theory.

2013. "The Relative Native." *HAU: Journal of Ethnographic Theory* 3, no. 3: 473–502.

2015. "Who Is Afraid of the Ontological Wolf? Some Comments on an Ongoing Anthropological Debate." *The Cambridge Journal of Anthropology* 33: 2–17.

von Lieven, Alexandra. 1999. "Divination in Ägypten." Altorientalische Forschungen 26: 77–126.

2000. *Der Himmel über Esna: Eine Fallstudie zur religiösen Astronomie in Ägypten*. Wiesbaden: Harrassowitz Verlag.

2006. *Grundriss Des Laufes Der Sterne: Das Sogenannte Nutbuch*. Copenhagen: Museum Tusculanum.

Von Staden, Heinrich. 1992. "Affinities and Elisions: Helen and Hellenocentrism." *Isis* 83: 578–95.

Wagner, Roy [1975] 1981. *The Invention of Culture*. Chicago: University of Chicago Press.

Wallis, Faith, and Robert Wisnovsky, eds. 2016. *Medieval Textual Cultures: Agents of Transmission, Translation and Transformation*. Berlin: De Gruyter.

Warren, James. 2004. "Ancient Atomists on the Plurality of Worlds." *Classical Quarterly* 54: 354–65.
Weidner, Ernst. 1967. *Gestirn-Darstellungen auf babylonischen Tontafeln.* Vienna: Hermann Böhlaus.
Weiher, E. von. 1993. *Spätbabylonische Texte aus Uruk. Vol.* IV. Mainz am Rhein: Gebr. Mann Verlag.
White, Hayden. 1978. *Tropics of Discourse: Essays in Cultural Criticism.* Baltimore, MD: Johns Hopkins University Press.
Wicksteed, P. H. and F. M. Cornford. 1957. *Aristotle Physics. Vol 1: Books 1–4.* Cambridge, MA: Harvard University Press.
Williams, Clemency. 2002. "Signs from the Sky, Signs from Earth: The Diviner's Manual Revisited." In *Under One Sky: Astronomy and Mathematics in the Ancient Near East*, edited by John M. Steele and Annette Imhausen, 473–485. Münster: Ugarit Verlag.
Williams, Raymond. [1976] 1983. *Keywords: A Vocabulary of Culture and Society*. Rev. ed. New York: Oxford University Press.
Winitzer, Abraham. 2017. *Early Mesopotamian Divination Literature: Its Organizational Framework and Generative and Paradigmatic Characteristics.* Leiden: Brill.
Winitzer, Abraham, ed. 2025. *The Spirit of Scholarship: Catholic Clergy and the Beginnings of Ancient Near Eastern Studies, 1875–1965*. Leiden: Brill.
Woolgar, Steve, and Javier Lezaun. 2013. "The Wrong Bin Bag: A Turn to Ontology in Science and Technology Studies?" *Social Studies of Science* 43: 321–40.
2015. "Missing the (Question) Mark? What *Is* a Turn to Ontology?" *Social Studies of Science* 45: 462–7.
Yano, Michio. 1994. *Abū Ma'shar: The Abbreviation of the Introduction to Astrology, together with the Medieval Latin Translation of Adelard of Bath*. With Charles Burnett and Keiji Yamamoto. Leiden: Brill.
1997. *Kūšyār Ibn Labbān's Introduction to Astrology*, Tokyo: Institute for the Study of Languages and Cultures of Asia and Africa.
2004. *Al-Qabīṣī: The Introduction to Astrology*. With Charles Burnett and Keiji Yamamoto. London: Warburg Institute.
2004. "Planet Worship in India." In *Studies in the History of the Exact Sciences in Honour of David Pingree*, edited by Charles Burnett, Jan P. Hogendijk, Kim Plofker, and Michio Yano, 331–48. Leiden: Brill.
2019. *Esoteric Buddhist Astrology: Japanese Sukuyōdō & Indian Astrology.* Translated by Bill Mak. New Delhi: Aditya Prakashan.
Yglesias, Elmer. 2000. "Greece: General Works." In *Reader's Guide to the History of Science*, edited by Arne Hessenbruch, 311–12. London: Routledge.

Index of Names and Subjects

For EU product safety concerns, contact us at Calle de José Abascal, 56–1°,
28003 Madrid, Spain or eugpsr@cambridge.org.

www.ingramcontent.com/pod-product-compliance
Lightning Source LLC
Chambersburg PA
CBHW051447150126
38315CB00002B/79
* 9 7 8 1 0 0 9 5 2 2 3 3 5 *